Ecological Studies

Analysis and Synthesis

Edited by
W. D. Billings, Durham (USA) F. Golley, Athens (USA)
O. L. Lange, Würzburg (FRG) J. S. Olson, Oak Ridge (USA)
H. Remmert, Marburg (FRG)

Volume 40

B. Nievergelt

Ibexes in an African Environment

Ecology and Social System of the Walia Ibex in the Simen Mountains, Ethiopia

With 40 Figures

Springer-Verlag Berlin Heidelberg New York 1981

Dr. Bernhard Nievergelt
Zoologisches Institut der Universität Zürich
Ethologie und Wildforschung
Birchstrasse 95, CH-8050 Zürich

QL
737
.U53
N53

Cover motive: A Walia ibex male of roughly 7 years in the mountain steppe of Simen with Giant Lobelias

ISBN 3-540-10592-1 Springer-Verlag Berlin Heidelberg New York
ISBN 0-387-10592-1 Springer-Verlag New York Heidelberg Berlin

Library of Congress Cataloging in Publication Data. Nievergelt, Bernhard. Ibexes in an African environment. (Ecological studies; v. 40) Bibliography: p. Includes index. 1. Bouquetin. 2. Mammals—Ethiopia—Simen Mountains. I. Title. II. Series. QL737. U53N53. 599.73′58. 81-5281. AACR2.

This work is subject to copyright. All rights are reserved, whether the whole or part of the material is concerned, specifically those of translation, reprinting, re-use of illustrations, broadcasting, reproduction by photocopying machine or similar means, and storage in data banks.

Under § 54 of the German Copyright Law where copies are made for other than private use, a fee is payable to "Verwertungsgesellschaft Wort" Munich.

© by Springer-Verlag Berlin · Heidelberg 1981
Printed in Germany

The use of registered names, trademarks, etc. in this publication does not imply, even in the absence of a specific statement, that such names are exempt from the relevant protective laws and regulations and therefore free for general use.

Typesetting, printing, and binding: Brühlsche Universitätsdruckerei, Giessen
2131/3130–543210

Acknowledgements

During the carrying out of field work and when completing this study, many persons and several institutions have given their support and I would like to express my sincere thanks to all of them. The project was mainly financed by the Swiss National Science Foundation (Project No. 4799.3). Additional aid was given by the Swiss Foundation for Alpine Research which has paid most travel expenses of the preliminary field trip and by the World Wildlife Fund which – apart from the backing and stimulating help of Dr. F. Vollmar, the General Secretary at that time – has provided a Land Rover that was given subsequently to the Ethiopian Government to complete the equipment of the Simen Mountains National Park. The Ethiopian Wildlife Conservation Department has initiated the study in 1966 and has given substantial aid, in particular the costs for workmen and material while building our house and the providing us with one assistant and three game guards. I am grateful to numerous persons connected with the Ethiopian Wildlife Conservation Department. I would like to mention Major Gizaw Gedligeorgis, the General Manager at that time, Ato Balcha Edossa, Mr. J. H. Blower, the Senior Game Warden in Addis Abeba, Mr. C. W. Nicol, who worked as engaged Park Warden for Simen from 1967 till 1969, and especially Ato Berhanu Asfaw, who, as assistant, shared much of the field work with us, took care of the daily measurements on temperature and precipitation and maintained occasional radio connection to the Wildlife Conservation Department in Addis Abeba. Not least, we like to mention the game guards: mainly Ato Hussein Hassan, Ato Ambaw, Ato Daud and Ato Amara who have joined us in the field alternately. All of them were enthusiastic observers. During a couple of months I have also received occasional assistance from a most animating peace corps volunteer, Mr. Montague Demment. I would also like to express my thanks to the subsequent General Managers of the Wildlife Conservation Department: Brig. General Mabratu Fessaha and more recently Ato Teshome Ashine who have given their personal interest and support to a number of further projects in Simen – financed by the Swiss National Science Foundation, the World Wildlife Fund and the Pro Semien Foundation (president: Dr. G. Sprecher, Chur) – that were considered of value in order to broaden the basis of this study.

Furthermore, we owe special thanks to Mr. and Mrs. F. Bernoulli from the Swiss Embassy in Addis Abeba, to Ato Ghiorghis Mellessa and Dr. Brown from the Health Center in Gonder and for much valuable advice and assistance to Dr. G. Kistler, Mr. E. Gehri and to my brother Mr. P. Nievergelt, Zurich. We are indebted also to ministries and various officials of the Ethiopian Government outside the Wildlife Conservation Department, in Addis Abeba,

Gonder and Debark for their good will, and we would like to express our profuse gratitude for their hospitality to all residents in the Simen mountains.

Prof. F. Hampel and Dr. W. Stahel, Zurich, have given valuable advice for the statistical treatment of the ecological data. In computing these data I was previously assisted by Dr. T. Hinderling; for the main analysis, including transformations of data and the running of numerous test programmes, Miss G. Hoffmann and Dr. H.U. Keller have been acting with tireless effort and much understanding for mathematically deficient biological data. The computing of data was carried out at the Calculating Center of Zurich University.

Prof. O. Hedberg, Uppsala and Mr. J.B. Gillett, Dr. P.J. Greenway, Mr. M.A. Hanid, Miss C.H.S. Kabuye from the East African Herbarium in Nairobi have kindly identified plant specimens and named our herbarium. Copies of the herbarium are kept at these two places, a third copy is stored in the Institute of Systematic Botany of Zurich University.

While completing this study I have received much valuable advice from Prof. E. Frei, Prof. F. Klötzli and Prof. M. Schüepp, Zurich, Prof. B. Messerli and Dr. P. Stähli, Berne, Dr. J.P. Müller, Chur, Prof. W. Müller, Marburg, and from Dr. M. Coe, Dr. J. Phillipson and Miss F.A. Street, Oxford. Dr. H. Hurni, Berne has given me additional data on climate as well as on people's farming methods, Miss H. Graf, Zurich, has assisted me in identifying fractions of foraged plants and Mr. R. Zingg, Zurich, has enlarged and recalculated my sample of ibex horn measures. Mr. C.A.W. Guggisberg, Nairobi and Mr. O. Schaerer, Zurich, have helped in my search for literature that covers the region of Simen.

I am most grateful to Mrs. P. Phillipson and Miss E. Ison, Oxford, for correcting my helvetic English, to Miss J. Stocker, Mr. R. Anderegg, Zurich, and Dr. H. Hurni, Berne, for carefully reading the whole manuscript, and to Prof. F. Klötzli, Prof. B. Messerli, Dr. H. Jungius, Gland, Prof. H. Kummer, Mr. R. Zingg and Mr. E. Stammbach, Zurich, for checking selected chapters. Their comments and criticisms were extremely valuable.

Miss C. Vischer has drawn the figures, Mrs. C. Ganz has typed the manuscript.

I am also indepted to the staff of Springer Verlag for their skill in getting this book into print, especially to Dr. D. Czeschlik for his interest and helpfulness, to Mrs. J. von dem Bussche for her care in checking the language and to Mrs. E. Schuhmacher for her patience.

The greatest part of my thanks goes to my wife Esther who shared the field work with me, did most of the plant collecting and was involved in numerous technical details arising in connection with a study of this nature. In this whole section, where I took the opportunity to express my gratitude to various institutions and many persons, the reader may have noted the alternate use of "I" and "we" and possibly has condemned it as grammatical instability. However, all the "we"s were put deliberately as the expression of the joint adventure of my wife and myself during this study in the unforgettable Simen mountains.

March, 1981   B. NIEVERGELT

Contents

The Social System of the Walia Ibex

Conservational Outlook

1 Introduction

The Caprinae are characteristic ungulates of palaearctic mountain regions. The Walia ibex, *Capra ibex walie*, a member of the Caprinae, has colonized successfully the Simen mountains in Ethiopia, and as such presents an interesting act in the history of the Caprinae. Apart from the question of how the Walia ibex or its ancestors ever reached the Simen mountains, the only place it occurs, I would like to stress a further point: about half of the existing ungulate species in the world are endemic to Africa. Eightyseven species of the Artiodactyla are reported to live in this continent, while there are only 17 in South America (Haltenorth 1963, Delany and Happold 1979). Such abundance of ungulate forms must have been attended by optimal utilization of the various habitats and form co-adaptation of sympatric species (Lamprey 1963, Bell 1970, Gentry 1970). Despite such intense competition however, the Walia ibex found its natural niche still open. Obviously, its adaptations to living on steep cliffs have favoured its establishment over the already existing ungulate fauna in an afroalpine mountain area. In fact, none of the Cervids has achieved a similar successful colonisation on the African continent, even though they are much older as a group, and have already shown themselves to be well established on the American and Eurasian continents. It is interesting to note that a serious attempt was made by British settlers in the last century to introduce the Red deer, *Cervus elaphus*, into Kenya. This attempt was a marked failure, as the animals released have disappeared in a very short time (pers. comm. J. Kingdon 7.6.1978).

In this ecological field study I shall concentrate on the geographical range of the Walia ibex. It contains a fascinating plant and animal community consisting of both African and Eurasian elements. Chapters 2 to 6 are an introduction to this unique afroalpine community. In determining ecology and social system of the Walia ibex, it is of interest to learn which aspects have been retained, and which have been changed by the African environment, giving rise to specific adaptations. These points are considered in the conclusions following the two major parts of this study: the analysis of the habitat and the approach to understanding the social system of the Walia ibex.

Whilst examining the niche and habitat of the Walia ibex, I have also studied those of the Klipspringer, *Oreotragus oreotragus*, and the Gelada baboon, *Theropithecus gelada*. The Klipspringer is in fact the species best adapted to mountain territory among those ungulates that have evolved within the Ethiopian region. Both the Klipspringer and the Gelada baboon inhabit the same area, are similar in size to the Walia ibex, and are relatively common, thus acting as its possible competitors. The relative significance of various environmental factors for these three mammal species when selecting their habitat is investigated. This analysis, mainly carried out with the help of a stepwise multiple regression, is the central element of this monograph and occupies Chapters 10 to 14. The method is described in Chapters 7 and 8. It leads to a theoretical distribution pattern that may be considered a habitat map of the Walia ibex and the Klipspringer. Comparative ecological data are also given for the Simen fox, *Simenia simensis*, the Golden jackal, *Canis aureus*, the Gureza, *Colobus abyssinicus*, and the Bushbuck, *Tragelaphus scriptus*.

In Chapters 15 and 16 an attempt is made to understand the social system of the Walia ibex, bearing in mind its specific ecological requirements, described in the above analysis. Other data concerning the reproductive cycle, rutting behaviour and group size of the Walia ibex are also used in the above context (Nievergelt 1970a, 1972a, 1974). For males and females, adults and juveniles of the Walia ibex, all such findings are considered as the framework of conditioning factors, within which inferences on the presumably optimal behaviour are presented. Based on such inferences a number of predictions are given and subsequently examined.

Within Chapter 4, where the Walia ibex is introduced, a separate section deals with the question of when it might have reached Ethiopia. The history of the climate and the early adaptation of the Caprini to bare mountains and to following the retreat of the glaciers are the basis of this consideration (Geist 1971a, Kock 1971, Schaller 1977).

When discussing ecological and sociological results, one would like to know whether an animal population is living under optimal conditions, its population density, and to which pressures, both natural and man-made, it is subject. The Walia ibex is one of the world's most endangered mammals and thus such data are also of particular practical importance. In the autumn of 1969 the main range of the Walia ibex was officially declared a National Park (Wildlife Conservation Regulations 1972, Blower 1970), and in 1978 UNESCO listed the Simen Mountains National Park as a "World Heritage Site" (World Heritage Committee Meeting in Washington, D.C., September 1978). Existing conservation measures are designed to try to eliminate two of the main threats to the Walia's survival, namely poaching and the destruction of its natural habitat. The test of whether these conservation measures are effective or not, will be the resulting stability, increase or decrease in the population size. Proposals concerning the monitoring of the Walia ibex population, presented to the Ethiopian Wildlife Conservation Organisation, were put into effect by the acting Park Warden and the Game Guards (Nievergelt 1971). Methods for estimating population size and changes and some results for the period from 1968 until 1977 are summarised in Chapter 9; threats to the ibex and its habitat, conservation measures and needs are discussed in the final Chapter 17.

In 1966 the Ethiopian Wildlife Conservation Department initiated investigations on the status of the Walia ibex. Prior to this Brown had noted their low number after a mission carried out for the IUCN in 1963. The alarming state of this endangered mammal was described in various publications and reports such as: Blower 1966, 1969; Brown 1965a, 1966, 1969; Gerster 1973; Hurni 1976; Kloetzli

1972, 1975b, c; Müller 1972, 1973a, b; Nicol 1972; Nievergelt 1969a, b, c, 1970b, 1972b, 1973; Vollmar 1969.

The data and results contained in this investigation are based on findings from a preliminary survey in November and December 1966, and, in particular, from field work carried out during the year February 1968 to February 1969. Additional data were collected during a short visit in 1971. Data on population fluctuations include reports based on observations made by the acting Park Wardens J.P. Müller (1971–1973), P. Stähli (1973–1975), and H. Hurni (1975–1977), assisted by the Co-Warden Ato Berhanu Asfaw. The analysis of the ecological distribution of the mammal species mentioned, particularly the theoretical distribution pattern, is based on the topographical map (Scale 1:25,000) made by Stähli and Zurbuchen (1978).

A Reader's Guide

While composing and writing this monograph I was thinking of two imaginary readers. One of them might be a responsible conservationist, a layman interested in African communities or a naturalist engaged in problems of nature protection. As the other reader I see the scientist; in view of the major subjects treated in this field study I am particularly thinking of an ecologist, a wildlife biologist or any expert familiar with the literature on the Caprinae and/or afroalpine communities. Very clearly, as the reading of the introduction already reveals, the various chapters of this monograph are not of the same interest to both imaginary readers. In the following, and for each section of the monograph separately, I shall therefore give a note on the necessity and value of the treated subjects for each of the two readers. For the sake of easy understanding I am calling them "conservationist" and "scientist". Of course, I am aware of the fact that the true reader will generally carry parts of each of the two pure types described above, however in diverse proportions. Thus, the two types refer to points of view rather than to real persons.

Chapters 2–6, p. 5–49, the Simen, an afroalpine community: This section that introduces the major elements of the Simen community is of a purely descriptive character. It is written in the conviction that the biology of a species has to be seen in the context of its environment, but also with regard to the unique situation and value of the area. The nature of this section is also documented by the amount of accompanying photographs. The *conservationist* will be particularly interested in these chapters with some probable exceptions such as the table and discussion on horn growth, which he may ignore. The *scientist* may skim through the headings and pages and pick out particular subjects only, such as the vegetation (Chap. 3), the morphological characteristics of the Walia (Chap. 4.1) or the history of climate (Chap. 4.2) or he may drop this whole descriptive and introductory section and consider it – with the help of the subject index – as a reference book during the reading of subsequent chapters.

Chapters 7/8, p. 51–76, methods and techniques on how data were obtained and analysed: Among the Chapters 7–14 that form the first major part of the study, these two chapters are essential for persons interested in how the results given in Chapters 10–14 were gained, with which restrictions the statistical calculations ought to be considered, and more generally for those readers concerned with methods. The *conservationist* may drop these two chapters with the exception of Chapter 7.1 and possible Chapter 7.2. The *scientist* will decide according to the headings of the chapters and his personal interest whether he may ignore some of this technical information.

Chapter 9, p. 77–81: This is a technical chapter on population estimates of the Walia ibex. It is presumed to be of interest to both readers, the *conservationist* and the *scientist*. However, it is not required for the understanding of the subsequent analysis of the ecological and social behaviour of the Walia ibex.

Chapters 10–14, p. 83–133, the niche and habitat of the Walia ibex, the Klipspringer and the Gelada baboon: This part of the study contains the data of the ecological analysis and includes also a habitat map of the Walia ibex and the Klipspringer.

The *conservationist* is invited to include this section in his reading but he may peruse more superficially some passages referring to the multivariate statistical procedure, such as in Chapters 10 and 13. However, I have tried to make the results understandable also to the interested layman, who is not familiar with the technical language. Applied ecological problems are touched for instance in the Chapters 11.7, 12 and 14. For the *scientist* this whole section will be of a primary interest.

Chapters 15/16, p. 135–169, an attempt to understand the social system of the Walia ibex: In this second major part of the study it is important for the reader to maintain the mental connection between the given inferences on the social behaviour of the various class members of the Walia ibex (Chap. 15.2), the subsequent predictions on the level of the data (Chap. 15.3) and the tests of these predictions (Chap. 16). This section is of minor importance for the pure *conservationist*, but – as the section above – it is of a primary interest to the *scientist*.

Chapter 17, p. 171–177: This chapter deals with the conservational outlook for this afroalpine community in the Simen mountains and is therefore of absolute importance to the *conservationist*. Nevertheless this applied chapter might be of some interest to the *scientist*.

The Simen, a Unique Afroalpine Community

2 The Simen Mountains: Geography and Climate of the Study Area

The mountainous nature of most parts of the country gave rise to the saying, "Ethiopia is the roof of Africa". In fact Ethiopia does not house Africa's highest mountains, but it does contain the largest extensions of alpine zones on the African continent.

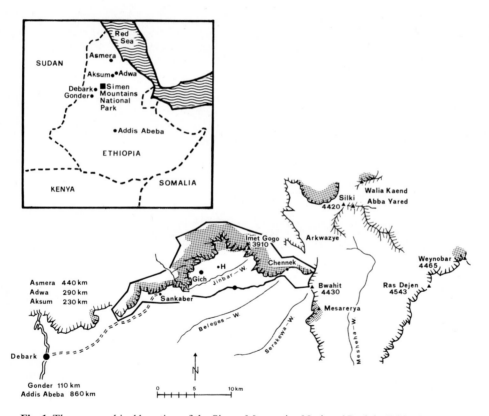

Fig. 1. The geographical location of the Simen Mountains National Park in Ethiopia (*above*) and within the Simen mountains (*below*). In the larger map the actual distribution of the Walia ibex is shown with *shaded patches*, the Park border with a *solid line* and the recently built Land Rover track leading from the highway in Debark to Sankaber with a *double dashed line*

Fig. 2. The topographically rich Simen escarpment, the main range of the Walia ibex. The small, white flowering shrub in the foreground is *Helichrysum citrispinum*, the tufted and large-leaved plant *Lobelia rhynchopetalum* (the Giant Lobelia). The view is shown from approximately point 3,786 south of Saha (map Stähli and Zurbuchen 1978) towards Northeast

A high basaltic plateau extends from Southern Ethiopia, where it is divided by the Rift Valley, to Northern Ethiopia and ends, approximately 100 km North of the ancient imperial capital, Gonder, in the Simen mountains. The highest points in the country are recorded in this mountain range, the highest of all being Ras Dejen at 4,543 m (altitude according to US Coast and Geodetic Survey 1961, see Stähli and Zurbuchen 1978). For geographical orientation see Fig. 1. However, the Simen mountains are less remarkable for their peaks than for the impressive escarpments

Fig. 3. In most parts of the Simen mountain range sheer cliffs clearly separate the plateau and the often rugged lowland areas. Here: Sheyno Sefer (see Stähli and Zurbuchen 1978), in the background Chennek and Bwahit

which separate the hilly plateaus from the terraced lowlands. Topographical features of the area include crests, gorges and terraces as well as separated rock towers; the so-called "Ambas" (see Figs. 2, 3 and 4).

For descriptions of the area from a general, topographical, geomorphological and/or geological point of view see Aerni 1978; Bailey 1932; Blower 1968a, 1970; Blumenthal 1962; Brown 1965a, b; Frei 1978; Gerster 1973, 1974; Hurni 1978; Klötzli 1975b; Maydon 1925; Mesfin Wolde Mariam 1970; Messerli 1975; Murphy 1968; Nicol 1972; Nievergelt 1969a; Rüppell 1838; Stähli 1978; Werdecker 1958, 1961. Outstanding maps of the area have been published by Werdecker (1967, scale

Fig. 4. The crests projecting from the *right* and the *left* are bounding a geographical unit "Geländekammer". In this case it is the area mainly overlooked from the observation points Kedadit and Muchila Afaf. The place of the photographer corresponds to the middle lookout of Muchila Afaf OP (see Fig. 23, p. 56), the lowest point is marked with *two arrows*. In the foreground, as well as middleground *Erica arborea*, the tree heather

1:50,000), more recently by Stähli and Zurbuchen (1978, scale 1:25,000), or are in preparation (Hurni 1980a in prep. scale 1:100,000). In Table 10 given later in a chapter on methods and comparability of data, the ecological and particularly the topographical ranges are given, with data on the relative number of variations of altitude, gradient of the slope and further criteria (p. 64). For a description of the study area see also Chapter 8.5 where – for statistical reasons – the relations between the selected environmental factors are presented and discussed (p. 71).

Two of the topographical features dictated, to a certain extent, the methodology and field technique. First of all, the study area was virtually divided into separate geographically defined units, "Geländekammern", bounded by the escarpment and the crests projecting from it (Fig. 4). Secondly, natural vantage points occurred along, or just below the escarpment, and on the crests, enabling the observer to overlook large areas of Walia ibex habitat. In fact, the topographical nature of the Simen mountains has provided natural, predetermined observation points.

The climate in the Simen mountains is of a typically afroalpine character, see Hedberg (1964), Coe (1967, 1969), Cloudsley-Thompson (1969), Lind and Morrison (1974) and Hurni and Stähli (in prep.). The mean temperature is relatively stable throughout the year, although some fluctuations particularly in the minimum temperatures are evident (Hurni and Stähli in prep.). However, variations in diurnal temperatures far exceed seasonal ones. This caused Hedberg (1964) to describe this phenomenon as "summer every day, winter every night". This impression is heightened in the daytime by the intense irradiation, the result of a thin atmosphere and – in the summer-half year – the near vertical position of the sun's rays. This produces high surface temperatures and a clear distinction between sun and shade temperatures. Another characteristic of this area is the fairly constant 12-h day, with a short twilight phase.

Seasonal changes are determined by the amount of precipitation. In the Simen mountains, the rainy season lasts from May to September/October, when precipitation usually is not too heavy, apart from 5 to 10 real hail-storms occurring mainly in the marginal months. The dry season occupies the rest of the year (see also Hurni and Stähli, in prep.). In Table 1 figures are given for temperature and precipitation over the year February 1968 to February 1969. In comparison with figures from subsequent years, the rainy season of 1968 was marginally less heavy. The total amount of precipitation in 1973 was 1,467 mm, and in 1974 1,591 mm, in 1975 1,548 mm, and in 1976 1,586 mm (measurements made by the respective Park Wardens P. Stähli and H. Hurni). The unexpectedly light rainy season of 1968 was partially compensated for by a short period of rains in November.

Observations on the Walia ibex were not disrupted by rain or fog, as our main observation times were during the early morning hours which were usually bright and cloudless. On most days it was not until 9 or 11 a.m. that clouds and fog arose within the escarpment and thus obscured the view (Fig. 5). Most often rain started at noon or later (see also Hurni and Stähli, in prep.). In heavier storms, the rain is mixed with hail, particularly at the beginning and end of the rainy season. I identified snow-fall only on one occasion: at the Silki-Pass at approximately 3,800 m on November 23rd 1968 (Fig. 6).

In Fig. 7 the direction and roughly estimated velocity of the wind are shown for four seasons. The figures are based on the sums of values made from observations of animals at the three observation points, Muchila Afaf, Kedadit and Saha (altitudes 3,400 to 3,785 m, see also Chap. 7.1). It is important to note that during the rainy season the most common wind direction – this is the case for stronger

Table 1. Precipitation and temperature for Gich-Camp, Swiss House (Park Center, 3,600 m, Coordinates 97,325/18,325 (map Stähli and Zurbuchen 1978) Lat. 13°16′ North Long. 38°6′ East, temperature was measured in the shade, 1 m above ground; and the monthly minimum and maximum temperatures for Muchila Afaf (3,360 m, Coord.: 95,430/19,150) and Imet Gogo

Location	Type of data		1969 Jan	68: 11.–29. 69: 1.–10. Feb	1968 Mar	19 A
Gich-Camp, Swiss House (3,600 m)	Precipitation	Total amount in mm	14	2	5	3
		Number of days with at least 1 mm	4	1	3	
	Temperature	Mean minimum (n nights meas.)	2.0 (31)	0.63 (26)	2.45 (31)	(1
		Mean maximum (n days meas.)	10.91 (31)	12.67 (27)	14.73 (31)	1 (1
		Absolute min.	0	− 2	0.5	
		Absolute max.	15	16	18	1
Muchila Afaf (3,360 m)	Temperature	Absolute min.	5.5	6.5	6	
		Absolute max.	14.5	14.5	19	1
Imet Gogo (3,890 m)	Temperature	Absolute min.	− 2.5	− 2.5	0.5	
		Absolute max.	11.5	12.5	13	1

Fig. 5. During the period of rains, usually in the late morning, clouds arise within the escarpment range and subsequently cover the whole area. The view is shown from the plateau in the Imet Gogo range to the North (25.9.1968)

(3,890 m, Coord.: 101,070/20,270). The thermometer at Muchila Afaf was placed in a tree heather approximately 1 m above ground and shaded with twigs and lichens. The thermometer at Imet Gogo was fixed in a shady rock crevice. The period of measurements lasted from February 11th 1968 until February 10th 1969

»8	1968	1968	1968	1968	1968	1968	1968	Year
ιy	June	July	Aug	Sep	Oct	Nov	Dec	
	221	392	184	136	55	97	3	1,349
5	20	29	18	19	12	8	2	137
3.89	4.83	5.08	5.08	4.66	3.08	1.91	1.66	3,51
1)	(30)	(31)	(30)	(30)	(31)	(30)	(31)	(350)
3.34	12.56	10.56	11.53	12.31	11.38	11.56	11.88	12.24
()	(30)	(31)	(30)	(30)	(31)	(30)	(31)	(351)
1.5	3	4	3.5	3	1	0	− 1	− 2
8	14.5	12	14	14.5	14	14.5	14	18
5.5	5.5	6	5.5	5	3.5	4.5	5	3.5
9.5	20	14	14.5	15	13.5	15.5	15.5	20
2.5	0.5	3	2 -	1	− 0.5	0	− 2.5	− 2.5
6.5	11.5	15	13.5	15	13	13	9.5	16.5

Fig. 6. Ras Dejen, with 4,543 m the highest peak of Ethiopia, on November 26th, namely 3 days after the snow-fall mentioned in the text. Ras Dejen is the summit to the *left*

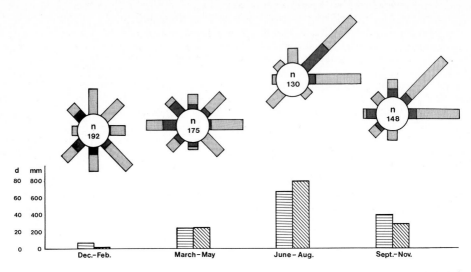

Fig. 7. Wind and precipitation in four different seasons in the Simen mountains. For the wind in the upper diagrams, the direction and velocity are shown, estimated in connection with observations of animals from the observation points Muchila Afaf, Kedadit and Saha. With *light grey*, weak wind velocity is indicated (class 1: light wind which keeps grass moving), *dark grey* was applied for stronger wind (as from class 2: stronger wind, branches of trees are twisted, upright walking is hindered). For the precipitation the number of days with rain of at least 1 mm is shown in the columns *left (hatched horizontally)* and the total amount of precipitation in mm in the columns right *(hatched diagonally)*. Measurements of precipitation were taken at the Gich house (see also Table 1)

wind in particular – is East and North-East. This result corresponds with the findings of H. Hurni who measured an average direction of precipitation over the year of 49° (in the 360° circle), which is roughly Northeast. The rain is inclined at an average angle of 76° (H. Hurni, pers. comm. 20.2.79). This means that slopes with a general easterly direction receive more rain than west-facing slopes.

3 The Vegetation in the Simen Mountains and Human Utilization of the Area

The three main vegetation belts distinguished in the literature are founded largely on the general appearance of the vegetation, as given by the dominating species. These fit in the following manner into the topographical features of the area as in (Hedberg 1955; Cloudsley-Thompson 1969; Coe and Foster 1972; Lind and Morrison 1974): (1) The upland forest belt or montane forest covers in its natural state the topographically rich lowland with wide terraces, ridges and gorges below 3,000 to 3,100 m. Among the tree species we find *Juniperus procera, Hagenia abyssinica, Olea chrysophylla, Syzygium guineense, Rapanea simensis*. As can be seen in Table 11 (p. 71), where various differences of gradients of the slope, of the vegetation and human influence are shown, there is much human activity in this belt. Several villages are located on the terraces and, as a consequence, only small areas from the natural forest are left. (2) The ericaceous or moorland belt has its lower limit at about 2,900 to 3,200 m and its upper limit between 3,600 and 4,000 m, with 3,700 m as mean for the Simen highland (Hurni 1980a, in prep.). It thus dominates the main part of the escarpment and is therefore less influenced by human interactions. Main trees and shrubs are the tree heather *Erica arborea*, and St. John's wort *Hypericum revolutum* (see Figs. 8, 9). In the lower section, among others *Nuxia congesta* and *Rapanea simensis* also join the heather belt. (3) The afroalpine belt consists of all the country above the timberline. This zone includes the higher terrain of the escarpment but extends mainly on the high plateau. It is characterised by grasses such as *Festuca macrophylla, Festuca abyssinica, Danthonia subulata* which are interspersed in some areas by Giant Lobelias *Lobelia rhynchopetalum*. At higher altitudes, the handsome but spiny shrub *Helichrysum citrispinum* is increasingly represented (see Figs. 2, 5, 6). In many parts of this belt, particularly in flat areas, domestic animals, mainly cows and sheep, are grazed quite extensively. However most eroded areas on the hilly plateau, such as the cultivated terrain around the village Gich are already located in the ericaceous belt (see Hurni 1978).

The three belts described are not of homogenous vegetation and are to be subdivided into different vegetation types. For this classification I shall follow Kloetzli (1975a and pers. comm.). Within the montane forest belt three types were distinguished. (a) The lower montane Cordia/Ficus/Pterolobium sclerophyll woodland: The treelayer is composed mainly of *Cordia abyssinica, Ficus vasta, Bersema*; the ravines contain *Pterolobium stellatum* and *Phoenix reclinata*. The following trees also occur in the other two montane forest types: *Euphorbia obovalifolia, Syzygium guineense, Schefflera abyssinica, Olea chrysophylla, Rhus*

Fig. 8. Tree heather, *Erica arborea* in the Set Derek-area. The lichen *Usnea* sp., well visible on the dead branches, is a regular "companion" of the tree heather growing within the escarpment range of the Park, but it is clearly less developed in areas with less humidity, such as the tree-heather belt on the plateau below the Gich camp

abyssinica, Rhus vulgaris. (b) the middle montane Syzygium/Maesa sclerophyll woodland: Syzygium dominates, while some of the above-mentioned species also occur as well as *Maesa lanceolata* and – in dryer sites – *Acacia abyssinica*. (c) The upper montane Olea/Maesa/Juniperus Sclerophyll woodland: The outstanding trees are mainly *Juniperus procera, Nuxia congesta, Rapanea simensis, Dombeya schimperiana, Hagenia abyssinica*, the latter two species grow preferably in the moister parts of the range. Particularly in the two upper types of the montane forest belt, but also in the lower parts of the subalpine Erica-woodland, shrubby species

Fig. 9. **a** Tree heather woodland, *Erica arborea* with *Usnea* sp, in the foreground *Lobelia rhynchopetalum*; **b** after the rains, *Kniphofia isoetifolia* is one of the plants that creates an impression of spring-time (altitude 3,300 m, original woodland opened up due to extensive human influence); **c** *Dianthus longiglomis* uses vertical or overhanging cliffs as habitat; **d** *Habenaria decorata*, a particularly attractive species among the various occurring orchids. **a–d** represent vegetation of the upper ericaceous belt; aspects and elements of the afroalpine vegetation are visible in various figures such as Figs. 2, 19, 22, 32, 40; of the upland forest belt in Figs. 10, 12, 20c

and creepers are abundant. I should like to mention *Myrsine africana, Rosa abyssinica, Echinops giganteum, Jasminum abyssinica* and *Clematis simensis.* In rocky and open sites *Aloë* sp., *Ferula communis* and *Kniphofia foliosa* represent attractive spots in the scenery.

Within the subalpine ericaceous belt, apart from some changes with altitude and with special habitats connected with relief and rock, Kloetzli (1975a) drew a distinction between two types, i.e., a drier and a more moist type of erica-woodland. The vegetation of the afroalpine belt, a puna-like mountain steppe, was subdivided by Kloetzli into four different types. As in the ericaceous belt, the distinction was drawn according to aridity, a peculiarity that is correlated with the depth of the soil. In Table 2 a list of some major plant species which occur in the two belts is given. In six columns for these two subalpine and four alpine vegetation types and in three additional columns for special rocky sites, stream-side habitats and pasture-land, it can be seen if a species occurs fairly regularly (×), if it is the or a main habitat of the species (× ×), or if the species is dominant (× × ×). Recorded in the last column

Table 2. List of some major plant species growing in the two aridity types of the Ericaceous belt (III and IV) and in the four types of the afroalpine belt (I–IV, classes according to Kloetzli 1975a). Type I is the most arid with only little soil; type IV represents the wettest habitat with well-developed soils. The 7th to the 9th columns stand for special habitats: below and above rocks (Rock), stream-side vegetation (Str) and pasture-land (Past). If a plant species occurs fairly regularly it is marked (×), if it is known to be the or a main habitat of the species (× ×), or if the species is dominant (× × ×). In the last column the flowering period is indicated by the number of the months as observed in 1968/69. Hyphens framing the figures mean that the beginning and the end of the flowering period is not known. As can be seen by a number of question marks and/or supplementary names, the proper names for several plants remain in doubt

Plant species	Ericaceous belt		Afroalpine mountain steppe				Rock	Str	Past	Flowering from–to (season 1968/69)
	III	IV	I	II	III	IV				
Trees and larger shrubs:										
Erica arborea L. (Ericaceae)	× × ×	× × ×								9–11
Hypericum revolutum Vahl (*H. lanceolatum* Lam.) (Hypericaceae)	× ×		× ×						×	11–4
Lobelia rhynchopetalum (Lobeliaceae)		×			×	× ×				7–3
Otostegia rependa Benth. (Labiatae)	×									–12–
Rapanea simensis (DC.) Mez. (Myrsinaceae)	×	×								
Rosa abyssinica R. Br. (Rosaceae)	×								×	10–3
Senecio sp. – Hedberg 4,160 (Compositae)					×		×			–10–
Small shrubs:										
Bartsia kilimandscharica Engl. (Scrophulariaceae)	×	×			×					–9–
Bartsia sp. (yellow flowers)	×	×			×					10–3
Bartsia petitiana Hemsley	× ×	×		×						8–3
Bartsia trixago L.	×								×	–9–11–
Blaeria spicata A. Rich (Ericaceae)	× ×	× ×			×					9–3

Table 2 (continued)

Plant species	Ericaceous belt		Afroalpine mountain steppe				Rock	Str	Past	Flowering from–to (season 1968/69)
	III	IV	I	II	III	IV				
Clematis simensis Fres. (Ranunculaceae)	×									–12–
Clematis sp. aff. C. grandiflora D. C.	×	×								–12–
Hebenstretia dentata L. (Selaginaceae)	×	×				×				7–12
Helichrysum citrispinum Del. var. citrispinum (Compositae)	×		×	×	×	×	×			12–3
Helichrysum cymosum (L.) Less. ssp. fruticosum (Forsk.) Hedb.					×	×	×			11–3
Helichrysum foetidum (L.) Cass.					×					–10–
Helichrysum formosissimum Sch. Bip. ex A. Rich.		×						×		–10–12–
Helichrysum gerberaefolium Sch. Bip.					×					–9–
Helichrysum splendidum (Thunb.) Lessing				×	×	×				11–3
Helichrysum horridum (Sch. Bip.) A. Rich.	×						×			12–5
Helichrysum sp = Hansson 199		×			×	×				–9–
Jasminum abyssinica DC. (Oleaceae)	×									–12–
Rumex nervosus Vahl (Polygonaceae)	×									–11/12–
Pteridophyta:										
Adiantum thalictroides Schlechtend.	×						×			
Asplenium adamsii Alston		×				×	×			
Cheilanthes farinosa (Forsk.) Kaulf.	×	×	×							
Gramineae/Cyperaceae/Juncaceae:										
Agrostis gracilifolia C. E. Hubb.			×	×						–10–
Agrostis kilimandscharica Mez.		×								–11/12–
Agrostis sp.				×	×	×				–10/11–
Aira caryophyllea L.	×		×	×	×	×				10
Andropogon amethystinus (A. distachyus L.)?	×			×	×	×				9–12
Andropogon abyssinicus Fres.	×								×	9–11–
Bromus pectinatus Thunb.	×	×			×					9–12
Carex fischeri K. Schum.								×		–7–
Carex monostachya A. Rich.		×			×	×				10/11
Danthonia subulata A. Rich.			×	×	×	×				10–12
Exotheca abyssinica (A. Rich.) Anderss.	×	×	×							9–2
Festuca abyssinica	×	×	×	×	×	×				9/10
Festuca macrophylla	×	×			×	×	×			10–12
Juncus capitatus Weig.	×							×		–9/10–
Koeleria convoluta A. Rich. (K. capensis (Steud.) Nees)?	×	×			×	×			×	9–12
Luzula abyssinica Parl.	×	×				×				8–10
Pentaschistis pictiglumis (Steud.) Pilg.	×	×	×	×	×	×				10–12
Poa leptoclada A. Rich.	×	×				×				–9/10–
Poa schimperiana		×		·		×				–9–
Poa simensis (Poa sp. cf. simensis A. Rich. = Hedberg 4,209)	×	×	×	×	×	×			×	8/9–
Scirpus setaceus L.	×						×			–9–
Vulpia bromoides (L.) S. F. Gray	×		×	×	×					9/10

Table 2 (continued)

Plant species	Ericaceous belt		Afroalpine mountain steppe				Rock	Str	Past	Flowering from–to (season 1968/69)
	III	IV	I	II	III	IV				
Iridaceae/Liliaceae:										
Albuca angolensis Bak.	×									–9/10–
Allium sp.									×	9/10
Aloë sp.	×						×			–5–12–?
Hesperantha petitiana (A. Rich.) Bak.	×			×	×	×				7–9
Kniphofia foliosa Hochst.	×	×					×			8–12
Kniphofia isoetifolia Hochst. ex A. Rich.	×	×		×	×					7–10
Merendera abyssinica A. Rich.	×		×	×	×		×			4–7
Romulea fischeri Pax	×	×	×	×	×	×				8–10
Scilla sp. (= Gebre-Selassie 650)	×									–1–
Orchidaceae:										
Disa sp.			×	×	×				×	7–9
Habenaria decorata A. Rich.	×	× ×								7–9
Satyrium schimperi A. Rich.	×									8–10
Commelinaceae:										
Commelina africana L.	×	×							×	7–9
Cyanotis barbata D. Don	×		×	×			×			7–10
Araceae:										
Arisaema enneaphyllum A. Rich.	×									5–8
Ranunculaceae:										
Delphinium dasycaulon Fres.	×									–9/12–
Ranunculus oligocarpus A. Rich.				×					×	–8/9–
Ranunculus oreophytus Del.				×					×	5–1
Crassulaceae:										
Crassula pentandra (Royle e Edgew) Schönl.	×		×	×	×		×			9–1
Crassula sp. (*alba*)	×	×		×			×			8–3?
Sempervivum chrysanthum Britten	×						×			–12–
Sempervivum sp.			×				×			3–5
Umbilicus botryoides A. Rich.	×	×					×	×		–9/10–
Saxifragaceae:										
Saxifraga hederifolia A. Rich. e descr.	×	×				×	×	×		3–9
Rosaceae:										
Alchemilla (abyssinica) rothii Oliv.	×	× ×			×	× ×		×		7/8, 10?
Alchemilla sp. (small)	×		×	×	×					8
Papilionaceae:										
Argyrolobium schimperanum A. Rich. e descr.	×								×	–9/10–?
Astragalus atropilosulus (Hochst.) Bunge var. *atropilosulus*	×	×			×					–10–
Lotus discolor E. Mey ssp. *discolor* (*L. tigreensis*, Bak.)	×								×	–9/10–
Medicago polymorpha L.	×								×	–9/10–
Trifolium acaule A. Rich.	×		×	×			×			–10–

Table 2 (continued)

Plant species	Ericaceous belt		Afroalpine mountain steppe				Rock	Str	Past	Flowering from–to (season (1968/69)
	III	IV	I	II	III	IV				
Trifolium arvense L.	×								×	–10/11–
Trifolium campestre Schreb.	×								×	–9–12–
Trifolium cryptopodium Steud ex. A. Rich.	×	×	×	×	×				×	–10/11–
Trifolium multinerve A. Rich. vergens ad. T. elgonense Gillett	×		×				×			–8–10–
Trifolium petitianum A. Rich.	×		×						×	9–12
Trifolium polystachyum Fres. Var. contractum Lanza		×						×		–8/9–
Trifolium simense Fres.	×		×	×	×					–8/9–
Vermifrux abyssinica (A. Rich.) Gillett	×								×	–10–
Onagraceae:										
Epilobium stereophyllum Fres.		×						×		–7–9–
Cruciferae:										
Arabis alpina L.	×	×			×	×	×			5–2
Arabidopsis thaliana (L.) Heynh.	×	×		×					×	–5–9–
Capsella bursa-pastoris L.	×								×	–9/10–
Cardamine hirsuta L.	×	×								8–10
Cardamine obliqua A. Rich.	×	×						×		–7–9–
Thlaspi alliaceum L.			×				×			–10/11–
Hypericaceae:										
Hypericum peplidifolium A. Rich.	×			×					×	–9–
Malvaceae:										
Malva verticillata L.	×								×	–10–12–
Geraniaceae:										
Erodium moschatum Ait.	×									–11/12–
Geranium ocellatum Jacq.		×			×	×	×			7–12
Linaceae:										
Radiola linoides Roth	×						×			–10–
Balsaminaceae:										
Impatiens abyssinica Hook. f.	×								×	9/10
Polygalaceae:										
Polygala erioptera DC.	×	×			×					12–5
Polygala sphenoptera Fres.	×								×	–12–
Umbelliferae:										
Anthriscus silvestris (L.) Hoffm.		×						×		–10–
Ferula communis L.	×									11/12
Haplosciadium abyssinicum Hochst. (Caucalis melantha (Hochst.) Hiern)?					×	×			×	–9–
Peucedanum winkleri Wolff		×						×		–10–
Pimpinella caffra D. Dietr.		×						×		–10–

Table 2 (continued)

Plant species	Ericaceous belt		Afroalpine montain steppe				Rock	Str	Past	Flowering from–to (season 1968/69)
	III	IV	I	II	III	IV				
Euphorbiaceae:										
Euphorbia sp.				×					×	–10–
Polygonaceae:										
Rumex abyssinicus Jacq.	×								×	–12–
Rumex steudelii A. Rich. e descr.	×								×	–10–12–
Caryophyllaceae:										
Cerastium octandrum Hochst. ex A. Rich.	×			×	×	×			×	7–10
occurs in 2 forms: sep. 4, pet. 4, st. 8	×			×	×				×	
sep. 5, pet. 5, st. 10	×				×	×				
Dianthus longiglumis Del.							×			1–4
Minuartia filifolia (Forsk.) Mattf.	×						×			9–12
Sagina abyssinica A. Rich.			×	×			×			4–10
Silene sp.					×	×				9–12
Silene flammulifolia A. Rich. e descr.			×	×	×					11–2
Uebelinia abyssinica Hochst.	×						×			8/9
Primulaceae:										
Anagallis arvensis L.									×	–10–
Primula verticillata Forsk. ssp. *simensis* (Jaub. & Spach) Smith & Fletch.							×			1–5
Gentianaceae:										
Swertia erythraeae Chiov.	×	×			×					8–12
Swertia sp. (white flowers)				×	×	×				7–10
Swertia kilimandscharica Engl. (light blue flowers)				×	×	×				7–11
Boraginaceae:										
Myosotis abyssinica Boiss & Reut.	×		×	×			×	×		5–12
Myosotis vestergrenii Strooh. flowers light blue, of "ordinary" forget-me-not type					×			×		7–12
Myosotis vestergrenii?/Cynoglossum? flowers deep blue, intens. perfumed					×		×			–8–
Labiatae:										
Nepeta azurea Benth.								×		–10–12–
Nepeta sp.	×		×	×			×			–10–12–
Salvia merjamie Forsk.	×			×	×	×			×	8–12
Satureja pseudosimensis Benth.	×	×	×	×	×	×			×	7–12
Satureja punctata (Benth.) Briq.	×	×	×	×						12–3
Thymus serrulatus Benth.	×	×	×	×	×		×			10–4
Scrophulariaceae:										
Celsia sp.	×	×						×		9–12
Veronica glandulosa Benth.	×	×		×	×				×	7–10–

Table 2 (continued)

Plant species	Ericaceous belt		Afroalpine mountain steppe				Rock	Str	Past	Flowering from–to (season 1968/69)
	III	IV	I	II	III	IV				
Plantaginaceae:										
Plantago afra L. var. *stricta* (Schollsb.) Verdc. (*P. psyllium* auct. non L.)	×		×	×					×	–9–
Plantago lanceolata L.	×								×	–10–
Rubiaceae:										
Galium aparine	×									–9–
Galium spurium L. var. *echinospermum* (Walbr.) Desf.	× ×	×						×		8–10
Valerianaceae:										
Valerianella microcarpa Lois.	× ×						×			8–10
Dipsacaceae:										
Dipsacus pinnatifidus A. Rich. see also *Simenia acaulis*	×									–12–
Scabiosa columbaria L.	×	× ×			×					8–3
Simenia acaulis (A. Rich.) Szabo e descr. similar to *Dipsacus pinnatifidus*, but without spines					× ×			×		7–11
Campanulaceae:										
Campanula edulis Forsk.	×									–9–1–
Wahlenbergia arabifolia (Engl.) Brehmer		×								–12–
Compositae:										
Anthemis tigréensis J. Gay ex A. Rich.				×						8–12
Aspilia pluriseta Schweinf. ssp. *pluriseta*	×									–9/10–
Bidens sp.	×		×	×			×			9/10
Carduus nyassanus (S. Moore) R. E. Fries	×	×				×		×		–10–
Coreopsis boraniana	×								×	–9–12–
Crepis newii Oliv. & Hiern	×								×	10–12
Crepis oliveriana (O. Ktze.) Jeffrey	×	×								8/9
Cotula abyssinica A. Rich.	×			×			×			8–11
Conyza stricta Willd. (= *Conyza* sp. = J. Reader Roetz 20?)	×								×	–10–12–
Conyza tigréensis Oliv. & Hiern	×			×						9/10
Dianthoseris sp.				×					×	–10/11–
Dichrocephala chrysanthemifolia (Blumel) D.C.	×	×		×			×			9–3
Filago spathulata Presl.	×						×			9/10
Gnaphalium undulatum L.	×									–10–
Haplocarpha rueppellii (Sch. Bip.) Beauv.				×				×		3–10
Launaea cornuta C. Jeffrey	×	× ×						×		9–3
Phagnalon nitidum Fres.	×									–11/12–
Plectocephalus varians (A. Rich.) C. Jeffrey	×								×	10
Senecio, various species										
Siegesbeckia abyssinica Oliv. & Hiern	×									–12–
Tolpis altissima Pers.	×	×								12–2

Fig. 10. The upland forest belt, the altitudinal range below 3,100 m is comparatively most intensively cultivated. Arising from settlements located on the lowland terraces, due to increased human population and depleted soils, agricultural activity is expanding towards the escarpment, and, regardless of the steepness of the slopes, forests are cleared and the land ploughed. View from Truwata, 2,860 m, a village on the lowland terrace below the escarpment of Shayno Sefer and East of Imet Gogo (2.12.1966)

months within which the plants were observed to be in flower. It will be necessary to refer to this information in order to identify in which phase the plant species were selected by the Walia ibex and the Klipspringer (Chap. 16.6, p. 165). A rough comparison of the various flowering periods shows that most plants blossom immediately after the rains. However, several species do not follow this rule, with the result that we can actually find some flowers in any season of the year.

Land use in Simen, cultivation, changes in settlements, traditions, religions with a distribution map of the settlements of Ethiopian Orthodox Christians, Moslems and Felashas (Ethiopian Jews) have been described by Stähli (1978). Information on the habits of the Simen farmers is also given in a detailed description of the market town Debark (Naegeli 1978).

A characteristic of the area and one of the basic problems in all efforts to preserve this unique ecosystem is certainly the increase in the human population and the fine network of settlements all over the Simen, with scattered villages even in extremely remote areas. Thanks to its character as an ideal area of refuge it provided a strong stimulus to clear, inhabit and cultivate places that were less and less accessible during the last century when interminable wars and ravages caused the people to retreat into more secure areas. For example there was the battle of

Fig. 11. On the plateau, above the timberline, the mountain steppe is used to graze cattle and sheep. In many areas, such as here between Gich village and the Gich camp of the Park, the destruction due to excessive feeding pressure is obvious. Within the tussock-grass meadow some Giant Lobelias can be seen

Dibil near the Bwahit-mountain in 1855 and the subsequent coronation of the victor as emperor Tewodros II of Ethiopia in Derasge (Stähli 1978). We can assume that the Italian invasion before the Second World War released a further wave of increase in the human population. Rapid growth in recent years has been documented by Hurni (1975b) and Stähli (1978). Arable land is gained by the burning of forests, the natural vegetation up to roughly 3,700 m (see above). Without forming terraces, the land is ploughed subsequently and almost regardless of the gradient of the slope. The land is cultivated roughly up to the timber line, the alpine meadows above are used to graze cattle and other domestic animals. The highest fields of barley are located at 3,800 m (Werdecker 1967), teff (*Eragrostis abyssinica*) and wheat are planted below the ericaceous belt. Disturbed types of Erica woodland and "puna" vegetation is described in Verfaillie (1978) and Kloetzli (1977), including the main derived pasture types.

The sharp increase in population and in addition the regular loss of soil due to careless farming methods have caused an obviously high pressure on land resources. The existing remnants of forests are rapidly being cleared and changed into arable land, and livestock feeds and tramples in excessive numbers on the meadows. As a consequence erosion follows, soils and montane agricultural potential have become exhausted and finally destroyed (Figs. 10–12). We must also

Fig. 12. The insidious destruction of the forest. The photographs show the lowland areas of Tiya (*terrace left*) and Truwata (*terrace right*) on the 28.11.1966 (**a**) and 15.1.1969 (**b**), respectively

realize that this process is not only catastrophic for Simen itself; due to the destruction of the water catchment areas there is also a delayed effect on the lowland areas (see for the problems of land use and resources also Dasmann et al. 1973; Danz 1975; De Vos 1975; Grimwood 1965; Hurni 1975a, 1978; Kloetzli 1975b; Mesfin Wolde Mariam 1972; Nievergelt 1972b; Winiger 1976).

As yet, some areas of this unique natural afroalpine community remain, but due to the widespread scattered settlements the original range has not only shrunk, it has become divided. Much of the so-far unaltered area was declared as the Simen Mountains National Park in 1969 (Wildlife Conservation Regulations 1972). However, it was not possible to separate natural habitats and cultivated country clearly. Within the park area there is also one village (Gich) and there are a number of other villages close to the park border. This fact illustrates the wide distribution of the settlements. In Table 11 (p. 71) it can be seen that the influence of man is particularly evident at altitudes of around 2,600 and also 3,400 m, where there is also a large number of almost flat fields. The lower altitudinal range corresponds to the lowland terraces, the upper range to the village Gich on the plateau (see also Stähli 1978, Fig. 5 and Hurni 1980a, b, in prep.). Thus man has not claimed all types of habitat in the Simen mountains equally. Ranges below the timberline that

Fig. 13. A drinking cup, made of a male horn of the Walia ibex. The whole cup is approximately 12–14 cm high. The owner proudly and carefully kept it, and expressly stated that it had to be handed down to his children and remain in the family

are also of a moderate gradient were selected first. In Table 11 we must remember that these data refer to the study area, where the influence of man is generally low; in most parts outside this range and outside the park border more and also steeper slopes are cultivated and are already devastated.

The Simen inhabitants are farmers, but besides the cultivation of land and grazing of domestic animals, the hunting of wild animals also has or had its place. The Walia ibex was hunted for meat, but the males also because of their horns which were used to make drinking cups. In Fig. 13 such a cup is shown. During our field study, but prior to the legislation of the National Park, the inhabitants knew that the Walia ibex was protected. There is good evidence that poaching activity was small during that time. More poaching was recorded at the beginning of 1971 (Nievergelt 1972b), but from autumn 1971 the improved control had its effect. I will refer to this section on human utilization of the area in the last chapter: the Simen, an ecosystem in danger (p. 173).

4 The Walia Ibex

4.1 The Walia Ibex: Its Taxonomy and Morphological Characteristics; Age Classes and Horn Growth

The Walia ibex was first described by Rüppell in 1835, after an expedition in 1832 which took him from Massaua, through Aksum and Adwa along the old trade route to Gonder, over Ataba valley, Silki Pass, Arkwazye, Bwahit, Inchetkab, Debark through the Simen mountains, the latter being the range of the Walia ibex (Fig. 1, p. 7, see also Rüppell 1838; Mertens 1949; Stähli 1978).

There is no general agreement on the taxonomy of the Caprini (Simpson 1945), and consequently the status of the Walia ibex remains in doubt. Most authors consider it to be a subspecies of the *Capra ibex* or the *Capra nubiana*. I have compiled a table containing the names of the authors and their propositions concerning the number of species in the genus *Capra*, and the status of the Walia ibex (Table 3).

The Walia ibex is a close relative of the Nubian ibex; I was therefore not surprised to find a bony process on the forehead of a skull also of the Nubian ibex (BM 1935.12.12.2 Sudan Sharak, Powell-Cotton, British Museum London), a characteristic described as a peculiarity of the Walia ibex (Rüppell 1835; Lydekker 1913; Parisi 1925; Harper 1945; Haltenorth 1963). Nonetheless the Walia ibex is different in certain characteristics from its northern neigbour (see below). The classifications that I prefer are (a) one which considers the Walia ibex, along with most other ibexes, as a separate species, or (b) one which considers the Alpine, Siberian, Nubian and Walia ibex, and possibly the West Caucasian ibex, as subspecies of the *Capra ibex*. This second solution is the one recently proposed by Schaller (1977).

Table 3. Taxonomic status of the Walia ibex according to the proposals of various authors

Author(s)	Date	No. of species in genus *Capra*	Status of Walia ibex
Kesper	1953	1	*Capra ibex walie*
Herre and Röhrs	1955	1	*Capra ibex walie*
Haltenorth	1963	4	*Capra ibex walie*
Schaller	1977	6	*Capra ibex walie*
Couturier	1962	1	*Capra aegagrus ibex nubiana walie*
Heptner et al.	1966	8	*Capra nubiana walie*
Lydekker	1913	9	*Capra walie*

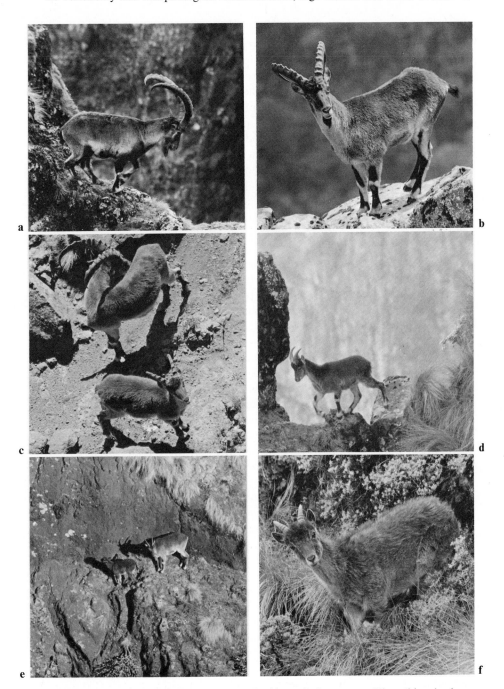

Fig. 14. The classes of the Walia ibex as set up in this study (see text, p. 30): **a** old male above 10 years, animal 1 in Table 32, p. 151 (class >7 years); **b** (and **c**, top animal) male of approximately 5–6 years (class >4–7 years); **c** lower animal, class: male of 3 to 4 years; **d** male in the second year (class <3 years); **e** adult female (animal on the right) with yearling (animal on the left); **f** young, above one year

Like most goats, the Walia ibex has a robust body. However, it is less massive and more sleek than the Alpine ibex, and it is stouter and heavier than the Nubian ibex. While reading the following description of the Walia ibex see also Fig. 14. Adult males are noticeably larger and heavier than females, although their most dominant characteristic is the long and heavy horns. The back and upper parts of the Walia ibex are chestnut-brown in colour. The chin, throat, and under side of the body, and the inner surfaces of the legs are whitish, with a black stripe extending down the front of each limb. A white band just above the knee cuts across the black stripe on the front legs. This black and white pattern is less pronounced in young males and females than it is in males of 3 to 4 years old and upwards. Males between 4 and 7 years have small black beards. Males of 7 years and older have considerably longer beards, and a darker chest. Females and young males do not have beards. A further characteristic, which is more pronounced in older males, is a black streak on the back.

During the fieldwork, the age of the animals was estimated as precisely as possible. For the treatment of data, however, in most cases the following sex-age-classes of the Walia ibex were set up: (1) Males older than 7 years of age; (2) Males of above 4, and up to 7 years; (3) Males of 3 to 4 years; (4) Males younger than 3 years; (5) Females; (6) Kids up to Yearlings; and (7) Young, from 1 to 2 years of age, whose sex is indeterminate as yet. For illustrations of individuals belonging to the various classes, see, apart from Fig. 14, also Figs. 4 and 5 and 7 to 10 in Nievergelt (1972a) and Figs. 42 and 44 in Nievergelt (1969a). For a photograph of a typical pair of Walia ibex horns see Parisi 1925. This shows the same features as can be found in Alpine and Siberian ibex horns (see Nievergelt 1978).

In Table 4 data are given that were taken from male horns of Walia ibex trophies available for measurement. For each criterion, the arithmetic mean, the standard deviation and the number of specimens examined are listed. The average values of each pair of horns were used as a standard. Thus, the number of specimens in general equals the number of pairs of horns. In exceptional cases, where only one horn sheath was preserved, the values of the single horn were considered the same as the averages of a pair. In the table, the corresponding values of measured horns of other ibex species and subspecies are given for comparative purposes.

In the right half of Table 4 the amount of horn growth per year is given. The lengths of the annual growth rings or segments were measured along the inner surface of the horns. In most Walia ibex trophies it was relatively more difficult than in other ibexes to take these measurements because as a rule the furrows between the annual growth rings were less pronounced and in some cases not clearly recognizable. Such horns or parts of horns had to be ignored for these data. This peculiarity of the Walia ibex must be explained by the afroalpine climate. In ibexes of temperate latitudes the break in the horn growth is obviously caused by the rut and the food shortage in winter (Duerst 1926; Nievergelt 1966a). But in the winterless climate of Simen one of the reasons for a break in growth is lacking. In fact out of four Walia horns where the date of death of the animals is known, the break in horn growth actually seems to occur in the rutting peak (March to May, see below, Chap. 4.3). According to the actual length of the last horn segment in comparison to the expected length if there were a full annual growth, the presumed growth breaks of these four horns were in February (two horns), March and May.

Table 4. Various horn growth measurements of male horns of five different *Capra* species and subspecies. For each measurement and Ibex form the arithmetic mean (m), the standard deviation (s) and the number of specimen used (n) is given. All values are written in cm. The data for this table were collected partly by Zingg (1980)

		Horn length at a chord of 30 cm	Chord at a horn-length of 30 cm	Circumference at 30 cm horn length	Diameter of horn at 30 cm horn length	Circumference after the 5th annual horn segment	Growth in length per year of age, measured along the knot-less back-side of the horn										
							Kid year	1	2	3	4	5	6	7	8	9	10
C. i. walie	m	37.5	25.4	19.3	7.9	24.5	11.0	9.2	8.5	9.2	10.0	8.6	6.0				
	s	2.7	1.6	0.8	0.2	–	1.8	2.0	2.1	1.8	2.8	1.7	–				
	n	8	5	5	2	1	4	4	5	6	5	4	1				
C. i. nubiana	m	42.6	23.6	13.0	4.8	15.3	10.9	10.4	10.1	8.6	8.9	7.0	7.3	5.7	3.5	2.7	2.3
	s	4.8	1.4	1.2	0.5	1.7	2.2	3.0	2.9	2.4	1.9	2.0	2.0	2.3	–	–	–
	n	24	22	22	14	10	16	18	20	19	16	14	9	5	2	1	1
C. i. ibex	m	33.3	27.2	18.4	6.8	21.0	7.5	8.8	7.5	7.4	7.2	6.9	6.5	5.6	4.8	4.1	3.5
	s	1.8	1.0	1.0	0.4	1.7	1.7	1.9	1.6	1.5	1.5	1.3	1.1	1.0	1.0	0.8	0.8
	n	23	114	112	102	336	140	393	410	388	375	351	339	307	258	212	179
C. i. sibirica	m	37.6	25.4	16.7	6.3	20.9	8.7	12.5	10.4	10.2	9.5	8.9	8.4	7.2	6.2	4.3	2.5
	s	2.4	1.1	1.1	0.5	1.7	2.3	1.9	2.2	2.4	1.3	1.2	1.2	1.0	1.6	–	–
	n	24	19	19	17	16	25	25	25	23	21	18	13	11	5	2	1
C. aegagrus (mainly from Iran)	m	40.7	24.0	15.5	5.2	20.1	10.4	12.4	10.8	11.0	11.0	10.0	8.9	7.5	6.2	4.0	4.2
	s	4.1	1.6	2.1	0.5	2.0	2.0	2.4	2.4	2.5	2.5	2.1	1.6	1.6	1.4	0.9	–
	n	8	12	10	113	113	83	152	146	134	114	86	57	38	21	6	3

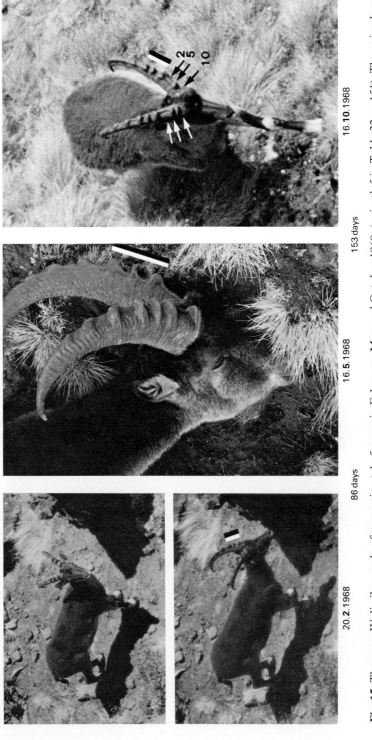

Fig. 15. The same Walia ibex male of approximately 5 years in February, May and October 1968 (animal 5 in Table 32, p. 151). The animal was identified in an analysis of the arrangement and form of the knots according to field drawings and various photographs. The most apparent sequence of three successive knots of different forms is indicated with a *black* and *white bar*. The horn basis at the three dates (2, 5, 10 respectively for the months of February, May, October) is indicated with *arrows* in the photograph to the *right* (see text p. 33)

Naturally these are only estimates based on an average age specific horn growth. Figure 15 shows the same Walia ibex male in February, May and October 1968. The two horn segments produced between the dates indicate in this case that the growth during the first period, that includes the rutting peak, was not reduced considerably. However, as in other members of the Capra ibex group, no knots were formed during the rut.

The values in the table indicate that the average annual horn growth of the Walia ibex is – at least in the first years of age – slightly greater than that of the Alpine ibex, but less than that of the Siberian ibex, the Nubian ibex and Wild goat. However, in contrast, the horns of the Nubian ibex and the Wild goat have a smaller circumference, (see columns 3 and 5). Columns 1 and 2 give values for the relationship between horn length and the corresponding chord length; both measurements are taken at the inner surface of the horn. The values give a measure of how much the horn tip is turned backwards. The chord length becomes smaller relative to the horn length the more the distal part of the horn has turned. These values show that the horns of the Walia ibex are more turned than those of the Alpine ibex and slightly less turned than those of the Nubian ibex and Wild goat. In column 4 the longer diameter of the horn cross cut at 30 cm horn length is given. This value compared with the corresponding circumference is a measurement of the horn form in its cross-cut. The relatively high values in diameter of the Walia ibex compared, for instance, with the Alpine ibex indicate the rather narrow rectangular form of the cross-cut of the Walia horn.

4.2 When in the History of the Climate Could the Ibex, as a Palaearctic Animal, Have Invaded the Range of the Simen Mountains?

Ibexes are virtually followers of glaciers and are adapted to bare mountains (Zittel 1891; Geist 1971a; Schaller 1977). Today, they are found in all larger mountainous areas of the Palaearctic region. During the last deglaciation period and with the beginning of vegetation succession, ibexes moved towards the deglaciated mountains and became separated by the beginning of forestation. Obviously, this geographical separation is a major reason for the evolution towards different species and subspecies. No other ungulate form has evolved and preserved the same valid and consistent adaptation to mountainous areas. A similar adaptation was attained by the sheep in North America, but in Europe and Asia, where they evolved together with ibexes, they did not compete successfully in rocky habitats (Geist 1971a).

Knowing these characteristics of the ibexes, the key to understanding the successful invasion of the Ethiopian region by this palaearctic mammal must be sought in the history of the climate in northern Ethiopia and North Africa, particularly in connection with the Ice ages.

Great climatic alterations took place in Africa during the Quaternary, which changed the face of this continent and influenced the biogeography. In the East African mountains particularly, many investigations based on pollen analysis were carried out and have led to a convincing concurrence on the climatic history of this region. In the following, I am relying on Butzer and Hansen 1968; Coetzee 1964;

Coetzee and Van Zinderen Bakker 1967, 1970; Farrand 1971; Gasse et al. (in press); Grove et al. 1975; Hamilton 1972, 1974, 1977, 1978; Kendall 1969; Kienholz and Messerli 1974; Livingstone 1967, 1971, 1975; Messerli 1975; Richardson and Richardson 1972; Street and Grove 1976; Street (in press, a, b); Van Zinderen Bakker 1962, 1964, 1967, 1969; Williams et al. 1978.

Several authors point out a close parallel in the temperature alterations occurring between the East African Region and the Northern Hemisphere during roughly the last 30,000 years. I should like to start from a cold interval which occurred between 26,000 and 14,000 years before the present. This period is equivalent to the end of the Würm, or Wisconsin, the last Ice age. This late glacial period was accompanied by increasing aridity which lasted approximately from 20,000 to 12,500/12,000 years BP. Before this late glacial period between 28,000 and 22,000 years ago, the climate in East Africa was less dry and possibly warmer. In the Mediterranean a pluvial phase is described which lasted from 50,000 until 20,000 years ago, while in the Sahara the climate seems to have been mainly moist before 30,000 years BP. The general decrease in humidity was possibly slower in the Northern Sahara than in East Africa. In the early postglacial period up to about 12,000 years BP the climate was dry in the area of Ethiopia with large numbers of grasses, herbs and composites in the pollen diagrams. It is presumed that the zones such as Podocarpus-Juniperus forest, Hagenia-Hypericum forest, Ericaceous belt and the afroalpine heath, as well as the snow line, have migrated upwards and downwards as a result of the Quaternary fluctuations in temperature and precipitation. In late glacial and early postglacial times the timberline may have been 1,000 m lower than today. In the Simen mountains the periglacial frost detritus belt with angular rubble on trough-shaped slopes due to solifluction processes was at least 700 m lower than the actual frost detritus belt, for which an average altitude of 4,225 m resulted (Hurni 1980a, in prep.). It is presumed that the Ethiopian montane forests lost a considerable number of species in dry periods such as that before 12,000 years BP, when it was much dryer than today, and that the modern flora and fauna of these forests consist partly of more recent immigrants. A major and important change from dry to moist climate took place between 12,000 and 9,000 years ago. This period lasted, with minor fluctuations, until about 5,000 years BP. The climate was definitely wetter, but probably less humid than in some other regions in East Africa, for instance Western Uganda, where a lush evergreen forest flourished. Since that time, the conditions have become dry again over wide areas.

In his brilliant glaciation theory Geist (1971a) stated that sheep evolved mainly during deglaciation periods from a possibly Ammotragus-like ancestor; that period for modern sheep for instance in North America must be assumed to be the post-Würm- (Wisconsin)-phase; thus, as from roughly 15,000 years BP. A parallel evolution seems probable for the Capra group, although – and this is in contrast to the sheep – the geographical distribution fits only sporadically with the morphological characteristics (Geist 1971a). However, this imperfect fit is understandable, if one considers that in the Ovis group the whole, postulated evolutionary lines were very clearly separated geographically, much more than those of the Capra group. The goats and ibexes evolved and are still distributed much more closely around the eastern Mediterranean, North Africa and the Middle

East, all of which are areas where ancestors may have occurred during the glaciation periods. Much of the evolutionary process involving the recent Capra members has obviously occurred in the upper Pleistocene. After the last Interglacial period (Eem) came the different phases of the last glaciation (Würm) beginning approximately 120,000 years BP when the situation was characterised by cold faunas. Zeuner (1945) writes: "By this time, immigration and adaptive evolution had supplied a large number of species well fitted for the periglacial biotopes".

Having in mind the specific adaptations of the ibex, it seems very likely that it extended its range from the northeastern corner of Africa towards Ethiopia in or shortly after the cold and arid late glacial period, but certainly prior to 12,000 years BP. In this period an open vegetation with depressed vegetation belts and a possibly weakened competition seem to have greatly increased the possibility of a successful ibex invasion towards the south. After the change of the climate between 12,000 and 9,000 years ago, forestation appears to have cut off rapidly the range of the invaded ibex from the north. Due to the topographical and climatic nature of the country it is very likely that the range of the Walia ibex was never extended much further south than it is today. Historical evidence for the Walia ibex to be a speciality of the Simen mountains can be seen in the fact that the Simen people had to offer Walias when paying homage to the king; e.g. for King Zar'a-Jacob, who ruled from 1434 until 1468 (Dillmann 1885). Thus the Walia ibex must have had a very limited distribution centuries before now. In connection with the period of invasion postulated for the Walia ibex, it seems noteworthy that those mammals, that are palaearctic in origin and are recorded for Ethiopia, are resistant to dry conditions (Corbet and Yalden 1972).

Reviewing the origin of the Caprinae and considering the apparently recent time the ibex has appeared in Ethiopia I would like to note a remarkable relation to a Christian narrative or legend about the way the Walia ibex came to Ethiopia. According to one version it was Abba (father) Yared, who brought the ibex from Palestine to Ethiopia (note, Dr. H. Hurni, Berne 1.2.1980). Abba Yared – it is said – in his older age lived as an anchorite in the Simen mountains. It is noteworthy that two neighbouring and high mountains in Simen are named: Walia Kaend and Abba Yared (see Fig. 1, p. 7). As stated by a different version, the ibex was introduced in the sixth century by the Nine Saints coming from Syria into this country (note, Dr. W. Raunig, Munich, 15.2.1980). There is a fascinating parallel between scientific evidence and legend in relation to the range the ancestors of the Walia ibex came from and the relatively recent time of their invasion. A further phenomenon revealed by this legend is that – as in other mountainous regions in Europe or Asia – the ibex apparently has a particular significance in the mental image of the native people (see for instance also Flury 1963; Huwyler 1977).

4.3 The Reproductive Cycle of the Walia Ibex and the General Grouping Pattern

I will refer to this chapter further below while discussing the social system of the Walia ibex, and thus, its lecture may be postponed up to p. 135. However, this information was placed here in this introductory chapter on the Walia ibex, feeling

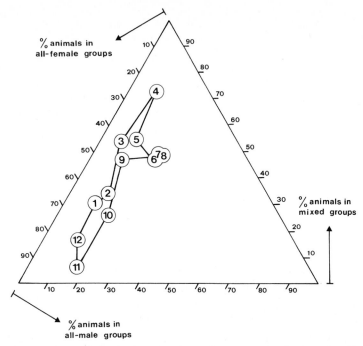

Fig. 16. Relative number of animals in all-male, all-female, and mixed groups for each month. The months are numbered from January to December. The absolute number of animals considered were from January to December: 186, 172, 200, 84, 221, 63, 117, 99, 208, 252, 64, 354

that it also improves the understanding of methods and results of the analysis of the Walia ibex habitat.

The Walia ibex shows rutting behaviour throughout the entire year, although there is an obvious peak from March to May. Older animals, both male and female, have a significantly more pronounced rutting peak than do younger ones. This phenomenon was described in earlier papers (Nievergelt 1970a, 1974) and I put forward the hypothesis that young growing ibexes can reach the onset of rutting behaviour at any time during the year. This stage of development is determined by age, physical condition and external stimuli. Older animals are more sensitive to these stimuli. It is not known whether the triggering stimuli are climatic, nutritional, social or of yet another nature. It is possible that the proximate factor controlling the timing of the reproductive cycle is the level of reserve protein. Jones and Ward (1976) have recently shown this in an important paper for Red-billed Queleas. This would be a plausible explanation for the reason why young ibexes are not yet adjusted to their reproductive cycles, as they do not have such a large accumulation of reserve protein. This hypothesis could also explain why in some ungulate species the reproductive cycle in both very young and very old animals is retarded (Cheatum and Morton 1946; Nievergelt 1966b, 1974) as well as in captive poorly fed animals (Nievergelt 1966b). This nutritional factor seems more likely to act as the proximate factor than the changing day length, which is fairly constant in the latitude of Simen (see p. 11).

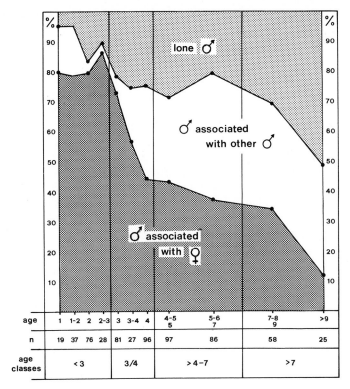

Fig. 17. The relative number of males of various ages in groups with females, in all-male groups and alone. In this figure eleven age classes are distinguished. The four classes of male considered normally in this study are given at the *bottom*

The reproductive cycle of a species is important with respect to its social behaviour. The general pattern of group structure for ibexes and sheep is one of all-male groups and of groups of females with young outside the rutting season. During the rut we have mixed groups (Nievergelt 1966a, 1967; Geist 1971a; Schaller 1977). In Fig. 16, in a triangle co-ordinate system, for the Walia ibex, the percentage of animals seen in single-sex groups and in mixed groups each month is shown. The relative number of animals in mixed groups is highest from March to May, the peak of the rut. Figure 16 also reveals the phenomenon mentioned earlier of continuous rutting behaviour throughout the year. The number of animals in mixed groups declines from May onwards with a trough in November, and then increases again towards the peak in April (see also Nievergelt 1974, p. 328).

Figure 17 shows the association pattern of males of different age classes. For each age group the number of males associated with females, with other males, and as lone males, is given. It can be seen that the number of males associated with females decreases with increasing age. A high percentage of males of medium age is associated with other males, and there are relatively more lone males with increasing age. I should also like to mention that the number of males associated with females declines rapidly after the age of 2 to 3 years. This age corresponds roughly to the age when males attain or even surpass the body weight of adult females. It also seems to indicate that the juveniles have reached maturity as young males, and start to join male groups.

Additional information on the group size and structure of the Walia ibex has been presented earlier (Nievergelt 1974). In the quoted paper, for the Walia and the Alpine ibex, it is also shown, that, on the average, mixed groups include more females than males (I refer to Fig. 4, p. 337). The association of polygyny with sexual dimorphism and environmental heterogenity coincides with the theoretical considerations on mating systems given by Orians (1969), Wilson (1975) and Ricklefs (1979, p. 272). A basic idea stems from the assumption that a heterogenous habitat leads to discrepancies in the value of the home ranges or territories. In such situations, females are supposed to do better in looking for a superior home range – with possibly other females being paired already – rather than to unmated males. Inferences on the social behaviour of the sex-age classes, results that refer to those inferences and a discussion of the data follow in Chapters 15 and 16 on the social organisation of the Walia ibex.

5 Two Possible Competitors of the Walia Ibex: the Klipspringer and the Gelada Baboon

The Klipspringer (*Oreotragus oreotragus*) is a small antelope, which partially shares the habitat of the Walia ibex (Fig. 18). Of all the ungulates that have evolved in the Ethiopian region, the Klipspringer is probably the species best adapted to steep slopes and rocks (Ferrar and Walker 1974). Aspects of the social organisation and ecology of the Klipspringer are given by Dunbar and Dunbar (1974a). They found the Klipspringer to be a territorial animal that forms a permanent pair bond, a grouping pattern that is in contrast to most forest duikers which normally occur singly. This behaviour of the Klipspringer is thought to be a possible adaptation to open habitats, where predation pressure is presumably greater than in covered areas. The male is seen frequently at look-out positions, possibly as protection for the female and the fawn.

A small group size is characteristic of other antelope species of similar weight and body size. It may have evolved due to the selective pressure of the high energy requirements of each animal (see Dorst and Dandelot 1970; Hendrichs and Hendrichs 1971; Tinley 1969, for papers on similar-sized animals also Jarman 1974; Wilson 1975). The higher per-gram metabolic requirement of small ungulates, compared with the lower demands of larger animals, gives rise to selective feeding behaviour, which consequently favours a small group (see also Schaller 1977). This tendency is augmented by the restricted possibilities of these small antelopes for energy storage. With this limitation we can expect the animals to be more sedentary, but this characteristic again does not permit large aggregations. The feeding behaviour of the Klipspringer seems to be more diverse than that of the Walia ibex, and it usually bites off only parts of individual plants (see Table 39, p. 165, as well as Wilson 1975).

In general my observations supported the findings of Dunbar and Dunbar (1974a; some of my observations were carried out in the same area as theirs). The following associations of Klipspringer were observed: 69 single animals, 104 two-animal groups, 60 three-animal groups, 9 four-animal groups, and 1 five-animal group. (The composition of the five-animal group was 1 male, 2 females and 2 offspring; 14.9.1968). The mating season is around August–September (according to Dunbar and Dunbar 1974a).

The newborn Klipspringer fawns show cryptic behaviour patterns, as do all other members of the Neotraginae and gazelles (see Hendrichs and Hendrichs 1971;

Fig. 18. A pair of Klipspringer, the horned male on the *right*, the hornless female on the *left*, watching the photographer with close attention (*above*); and a hiding young Klipspringer (see text p. 42)

Fig. 19. Adult Gelada male (*above*) and a part of a herd (Ohsawa 1979) feeding intensively in the afroalpine mountain steppe on the plateau (3,790 m). Geladas are endemic to Ethiopia

Walther 1968). On one occasion I approached a young Klipspringer which was lying down, and as I came closer, it hid (Fig. 18). All the time, I was watched closely both by the fawn, and by its parents who were standing at a distance of approximately 60 m.

The findings of Dunbar and Dunbar, along with my own observations indicate that the Klipspringer's average reproductive cycle is retarded, when compared with that of the Walia ibex. In Sankaber the Dunbar's saw the first fawns in November and December; the main parturition period for the Walia ibex is September/October (see Nievergelt 1970a, 1974). I presume this apparent difference in the fawning season must be related to the habitat occupied at that time and the state of the vegetation available. According to my observations, the characteristic range selected by Walia ibex females and their offspring in this season was steep and extremely inaccessible cliffs around rock pinnacles as described in the paper quoted above. These places provide excellent protection against the larger carnivores such as Leopard and Hyena; but as the thin soil in such exposed terrain is not able to retain much water in dry weather, the vegetation has a short green phase only after the rains. These slopes therefore quickly turn yellow in the course of the dry season. The Klipspringer in contrast does not frequent such extreme habitats. Further considerations on energy requirements, body size and condition of food are given in Chapter 16.6. Apart from explanations associated with nutrition, it is also possible that the shorter growing period of the young Klipspringers provides an additional reason for the retarded fawning season.

Another potential competitor of the Walia ibex is the Gelada baboon (*Theropithecus gelada*), a primate endemic to Ethiopia, and adapted to mountainous habitats (Fig. 19). Over the last few years, this almost purely vegetarian primate has been the subject of many studies, which include behaviour and ecology in the wild, such as Crook (1966), Dunbar (1977a, b, 1978b), Dunbar and Dunbar (1974b, c), Kawai (1979) and behaviour in captivity, for example Fedigan (1972) and Kummer (1974).

In the Simen mountains, Geladas can be observed in aggregations of very different sizes. Crook (1966) and Ohsawa (1979) describe the one-male unit or the one-male group as the fundamental unit of the Gelada society. Such a unit has an average size of around ten animals. Several one-male units which are joined by all-male units may aggregate to a herd, but there are also independent one-male units and all-male groups. For the herd, Dunbar (1977a) uses the term band which consists of a number of one-male reproductive units and 1–3 all-male bachelor groups. Ecologically important is apparently the subdivision in constant and relatively small groups. Dunbar (1977a) names the band as being the basic unit of the population. The herd (Ohsawa) or band (Dunbar) is not a closed association with strict membership, as one-male units may enter and leave. In addition, two neighbouring herds may aggregate to a multi-herd (Ohsawa) or combined herd (Dunbar). There are indications that the size of herds or multi-herds is correlated with the availability of food, in that abundant food permits the aggregation of larger foraging herds (Crook 1966; Dunbar 1977a). However, there are also observations which do not coincide with this interpretation. Ohsawa has counted the largest multi-herd as being composed of more than 620 individuals at the end of the dry season. Obviously, as with other mammals, the size of the groups is influenced by dozens of different pressures (Bertram 1978).

In contrast to the Walia ibex and the Klipspringer, the Gelada baboons are not shy at all. This phenomenon indicates that they are not actually harmed by people.

The Klipspringer and the Gelada baboon are included in this ecological field study in order to show their specific habitat selection, and then compare it with that of the Walia ibex. The results of this comparison are given in Chapters 10 to 12.

6 Further Mammals and Birds Living in the Simen Mountains

In the following pages I shall list first the mammal species which we could observe or which are reported to range within the study area in Simen. A selection of remarkable birds is given subsequently. The lists are given in order to demonstrate the living environment of the Walia ibex, the Klipspringer and the Gelada baboon. For the mammal species, apart from our own observations, I refer especially to various observations made by Dr. J.P. Müller, Chur, during his time as park warden in the Simen Mountains National Park from 1971 to 1973 (Müller 1973b, 1977). For each species, some remarks on the specific habitat are given. The animals are listed in the sequence of the usual taxonomic order. Those species marked with an asterisk are considered together with the Walia ibex, the Klipspringer and the Gelada baboon in an ecological comparison in Chapter 12.

Small mammals: *Arvicanthis abyssinicus* (Fig. 20b) is the most frequent small mammal in the grassland of the alpine belt, and it occurs in a density of 65 to 250 animals per hectare (Müller 1977). As an animal active each day it is a major prey species of the Simen fox, and the Golden jackal, as well as of birds of prey such as the Augur buzzard and the Tawny eagle. For further small mammal species see Müller (1977).

Crested Porcupine, *Hystrix cristata*; Quills were found in the Mayschaha valley and, according to Dunbar and Dunbar (1974d), at Sankaber.

Hamadryas baboon, *Papio hamadryas*; Müller (1973b) reported a herd of around 40 animals living on the southern slopes of Sankaber towards Addis Gey. There is no evidence for Hamadryas in the northern lowlands (see also Dunbar and Dunbar 1974b).

Anubis baboon, *Papio anubis*; several groups of around 20 animals were observed by Müller (1973b) in the montane zone (lowlands) near Adermas, Shagne, Muchila, Agidamia and Dirni; in this last area a group of Anubis was observed on one occasion mixed with 50 Geladas (see also Dunbar and Dunbar 1974b).

Grivet monkey, *Cercopithecus aethiops*; they were seen by Müller (1973b) mainly in the Acacia woodland (lowlands) near Antola, Muchila and Adermas.

*Gureza, Abyssinian black- and white Colobus, *Colobus abyssinicus* (Fig. 20c) single animals, but mainly small troops were seen occasionally and heard, when they were moving, jumping and roaring in the montane forest. Gureza also ranged although less frequently, in the lower ericaceous belt (see Marler 1969). My observations coincide with those of Müller (1973), who also spotted the Gurezas mainly in the western lowlands of the study area around Adermas and Muchila.

Fig. 20. Four remarkable mammals occurring also in the Simen mountains: **a** the endemic and endangered Simen fox, *Simenia simensis,* **b** its main prey, the Grass rat *Arvicanthis abyssinicus,* also possibly endemic to Ethiopia (Müller 1977); **c** a primate of closed forests: *Colobus abyssinicus* and **d** rarely observed, but ranging up to the highest altitudes, the Serval *Leptailurus serval*

*Golden or common jackal, *Canis aureus*; this species occurs throughout the whole area. It was seen or heard often on the high plateau near the house at all times of the day. Sometimes it was seen around garbage places and on carrion.

*Simen fox, *Simenia (ev. Canis) simensis* (Fig. 20a); the Simen fox is a carnivore endemic to Ethiopia; the animal occurring in Simen is described as the northern subspecies. We observed the fox mainly in the open grassland. Its main prey is obviously small rodents. In 1966, 1968/69 and 1971 I saw only lone animals, but in four observations the fox was associated with a jackal. Müller (1973b) out of 38 observations, saw two animals together on eight occasions, but he had no definite evidence for young either. The northern Simen fox – assuming the differentiation into subspecies is justified – is certainly one of the most endangered mammals of Africa (see Müller 1974).

Spotted hyena, *Crocuta crocuta*; this widespread species is also relatively common in Simen and is heard very regularly at night. In Simen, their main food is apparently domestic animals which occur in excessive numbers and are often tended with little care (see also the remarks to the vultures given below).

Serval, *Leptailurus serval* (Fig. 20d); I have observed the Serval only twice, once at the edge of the escarpment between Saha and Imet Gogo (Fig. 20) and a second time east of Ras Dejen.

Leopard, *Panthera pardus*; during our field period, one leopard was observed by M. Demment at Set Derek. The Leopard occurs in Simen, but the density seems to be very low (Myers 1973).

These carnivores are not likely to be real threat to the Walia ibex. The Leopard and the Serval are extremely rare, they were both heavily hunted. The Golden jackal and the Simen fox are not effective as enemies of the Walia ibex. However, it is possible that the Jackal may be an occasional predator of the Klipspringer and the Duiker (see Schaller 1972). The Spotted hyena – in Simen active during the night only – most likely does not occur in the inaccessible ibex terrain.

Rock hyrax, *Procavia capensis*; Müller (1973b) detected three colonies in the Simen Mountains National Park: near Chennek, above Muchila and near Sankorafa.

Bush-pig, *Potamochoerus porcus*; the species seems to occur in the montane zone. The only observation made by myself obviously refers to a rather untypical habitat. On 13.11.1968 one animal was moving just below the Muchila Afaf observation point at 3,330 m on a small and rather steep terrace with open vegetation; below and above were vertical rocks.

*Bushbuck, *Tragelaphus scriptus*; the species occurs as a small population in the montane forest. Coinciding with my observations Müller (1973b) reports as areas: Muchila and Adermas.

Grimm's duiker, *Silvicapra grimmia*; the Duiker occurs in the Heather forests around the Djinn Bahr River as well as in the forested parts of the Sankaber areas (see also Dunbar 1978a; Müller 1973b).

Checklists of the Simen birds have been presented by Lilyestrom (1974), Dunbar and Dunbar (1974d) and Bosmans and Moreaux (1977). However I should like to mention some specially characteristic birds of the area. Two species are remarkable because of their spectacular nose-dive in front of the escarpment: the White-collared Pigeon and the Chough. But as group the omnipresent birds of prey must be mentioned first. As a possible adaptation to the numerous birds of prey, in an earlier paper, I have put forward the hypothesis that females of the Walia ibex might co-operate in guarding and defending of kids against these birds. The hypothesis bases on the unusual behaviour of the females to remain in groups during parturition (see Nievergelt 1974, pp. 338/339). The species are listed using the numbers given by Urban and Brown (1971):

45. Wattled ibis, *Bostrychia carunculata* (Fig. 22a); common on the high plateau.

84. Black kite, *Milvus migrans*; common, particularly around villages.

86. Egyptian vulture *Neophron percnopterus* (Fig. 21b); occasionally, not shy.

87. Bearded vulture, Lammergeier, *Gypaetus barbatus* (Fig. 21a); common, picks up bones and lets them drop and break on rocks. On one occasion, I observed a Bearded vulture picking up the same bone four times.

Fig. 21. Birds of prey are omnipresent in the Simen mountains. **a** the Bearded vulture *Gypaetus barbatus*, almost a component of the Simen scenery; **b** the Egyptian vulture *Neophron percnopterus*, a regular visitor in human settlements; and **c** Rüppell's Griffon vulture, *Gyps rüppellii*. The noticeable abundance of birds of prey, and vultures particularly, is apparently maintained at this high level by the numerous domestic animals

Fig. 22

90. Rüppell's Griffon vulture, *Gyps rüppellii* (Fig. 21c); common, regularly on carrion.

92. Lappet-faced vulture, *Torgos tracheliotus*; common. All three species (87, 90, 92) are apparently largely dependent on the availability of carcasses (Kruuk 1967). In Simen, the main food resources for these vultures are domestic animals such as sheep, goats, cows and donkeys having stumbled on cliffy terrain or being killed by a hyena or another predator. The density of wild ungulates is most likely too low to support the observed abundance of vultures.

119. Augur buzzard, *Buteo rufofuscus* (Fig. 22g, h); common, occurs in a white and melanistic phase.

122. Tawny eagle, *Aquila rapax*; common.

148. Peregrine falcon, *Falco peregrinus* (Fig. 22e); occasionally, particularly in cliffy areas.

159. Erckel's francolin, *Francolinus erckelii*; observed occasionally in the tree-heather.

293. White-collared pigeon, *Columba albitorques* (Fig. 22d); this endemic species performs daily altitudinal movements, feeding on the plateau during the day, shooting down the rocks in the evening and roosting probably in caves at lower levels. In the morning, they can be seen spiralling their way slowly up (Brown 1965a; Boswall and Demment 1970).

297. Pink-breasted dove, *Streptopelia lugens*; common, not shy.

539. Hill chat, *Cercomela sordida*; common, not shy

795. Slender-billed chestnut wing Starling, *Onychognathus tenuirostris* (Fig. 22c); particularly on Giant Lobelias.

821. Chough, *Pyrrhocorax pyrrhocorax* (Fig. 22a, b); remarkable as palaearctic element.

827. Thick-billed raven, *Corvus crassirostris* (Fig 22f); endemic, common.

In Chapters 2 to 6, the reader was introduced to the afroalpine community in the Simen mountains. The first major part of this study now following includes Chapter 7 to 14 and contains mainly an ecological analysis of the habitats of the Walia ibex, the Klipspringer and the Gelada baboon. Within this part, the first three chapters (7 to 9, pp. 51 to 81) are of a technical character. Therein methods of obtaining data, methods on the analysis of data and also estimates of the Walia ibex population are given.

Fig. 22. A number of birds characteristic to the area. **a** *left* Wattled ibis, *Bostrychia carunculata;* **a** *right* and **b** Chough *Pyrrhocorax pyrrhocorax;* **c** Slender-billed chestnut wing starling, *Onychognathus tenuirostris* at an inflorescence of the Giant Lobelia; **d** White-collared pigeon, *Columba albitorques;* **e** Peregrine falcon, *Falco peregrinus;* **f** Thick-billed raven, *Corvus crassirostris;* **g h** Augur buzzard, *Buteo rufofuscus*, the photograph **g**, showing the nest in a tree heather, and, **h** the buzzard in a typical lookout position on an inflorescence of the Giant Lobelia

Methods, Techniques and Population Estimates

7 Methods of Obtaining Data and Field Techniques

7.1 Observation and Recording of Animals

Methods of observation were designed to give priority to ecological questions such as the habitat preference, distribution pattern, population size and density, age distribution, and birth rate of the Walia ibex. Information on feeding habits, social behaviour and group patterning was also collected.

In fact, during my preliminary survey it became clear that the selected methods of observation would, for the most part, be governed by the topographical peculiarities of the Walia ibex's ranges (see p. 11). The terrain included gently sloping plateau, deep steep-sided valleys, crests, gorges, terraces, and varying density of vegetation. This often meant that gaining access to various parts of the range was either time-consuming or limited, with differing degrees of visibility in these areas. Methods that have been successful elsewhere, e.g., counting big game, cannot be applied under these conditions. Furthermore, it was more or less impossible to follow systematically a group of Walia ibex all day long. Additional limitations placed upon our observations arose from the fact that the Walia ibex is an endangered species of mammal. As such our actual number of observations was expected to be low and we could not justify the trapping of individuals in order to mark them.

However, the topographical features of the area do have their advantages, such as the natural observation points overlooking the Walia ibex range, and the range itself is divided into separate well-defined geographical units. Observations made overhead on an area are advantageous because they allow the observer to record more easily and precisely the relative positions of, and the distances between, the various animals in a group. The observer is also able to spot the animals that are on ledges or terraces which they tend to frequent.

Within the main study area there were nine observation points (OP) which were visited regularly, the order guided by a general programme. Modifications due to other programmes or external needs such as collecting plants, buying food or due to weather conditions were tolerated. However, the degree of flexibility was limited by a minimum programme defined for each of the OP's. There were six other observation points further to the east and west of the main study area, and these were used occasionally. All the observation points were situated on, or close to the edge of the escarpment, and overlooked large areas of Walia ibex habitat, usually the major part of the geographical unit above which they were situated. If the hoped-for wide angle view was narrowed by rocks, trees or other close obstacles, an observation point was divided up into several lookout positions which were visited one after the other. Animals seen at the observation point itself, as the observer approached it, were also recorded. This extension of the observation area was made so that every animal visible from the OP was recorded. Figure 23 shows the location of the nine observation points, and their lookouts, in the main study area. The two lookout stations at observation point Kedadit (G4) were two separate observation points originally, and are exceptional because they were visited alternately. The implications of this exception are considered below. Figure 23 also shows the location of the Gich camp with our house. Each of the nine observation points could be reached from there, and it took between 30 and 90 min to walk to a point. Even though Walia ibex and Klipspringer could not be seen from the house, in some cases it was still considered an observation point because both the Simen fox and the Golden jackal were seen nearby.

Tables 5, 6 and 7 refer to the observation points in the study area. All three tables include all of the observation points. Table 5 gives the number of visits made to each observation point for each month and for the whole year. Visits were not included if there was heavy fog or if the time spent at an observation point was less than 30 min. The total spotting time was 367 ¼ h over the whole year, and thus on average 1 ¾ h per visit. Implications concerning spotting time are discussed below. Table 6 gives the number of animals of each species recorded at each observation point. Obviously, the Walia ibex, the Klipspringer and the Gelada baboon are the quantitatively dominate species of larger mammals within the observed areas. Comparisons between the recorded numbers of the various species must be made with caution because the observation points were selected to cover primarily the distribution zones of the Walia ibex. The low recorded numbers of the Gureza, the Bushbuck and the Bush-pig can be explained by the fact that these species favour vegetation zones at lower altitudes, the Duiker apparently avoids open vegetation and cliffy areas and the Simen fox, the Golden jackal, the Serval and the Leopard are all carnivorous and so have as such a relatively low density. The results at the Gich camp observation point (H), a site not located at the cliff edge, clearly deviate from those at the regular observation points. It must be noted that on the long excursions east of the main study area Klipspringer and Gelada baboons were ignored. Two such excursions were carried out in areas east of Chennek and Bwahit, one in 1966 into the Silki area and one in 1968 towards Silki, Abba Yared, Walia Kaend, Ras Dejen and Weynobar. These excursions were carried out in order to examine further places of possible and earlier distribution of the Walia ibex.

Table 7 contains data on Walia ibex observations only. The table shows particularly how many Walias of the distinguished sex- and age classes were observed from each observation point. These data are given in the columns M and (F) and M >7 to F at two levels of differentiation. In the first two columns we can find the number of visits and the total number of Walias. For several OP's this figure is higher than the addition of M and (F) and this value is higher than or equal to the sum of M >7 to F, as not each spotted individual could be classified reliably at the same level of differentiation. The third column lists the codified number of Walias as explained later (p. 68) and the fourth the number of observations, thus the number of observed groups or lone animals. An animal was defined as associated with a group if it was within 100 m of its nearest neighbour. In column five we find the average group size and in the last two columns two fertility rates are given.

From the three tables 5 to 7, it can be seen that special attention was paid to the observation points Muchila Afaf (G3) and Kedadit (G4), which overlook the same geographical unit but from a different position, and to Saha (E2), a prominent crest which juts out the adjoining geographical unit to the east (see Fig. 23).

I visited the observation points with my wife and/or Ato Berhanu Asfaw and one or more of the game guards. Both binoculars and a telescope (Kern Alpitrix; 15, 28 and 45 enlargement) were used on these excursions. If animals were seen, the following information was recorded, separately for each group and in this sequence: (1) group structure and the estimated distance between individuals. This was not done for the Gelada baboon where I simply counted or estimated the number of individuals; (2) behaviour, i.e., the activity of the animals at the recording time; (3) the age and sex of each minimal (see p. 30), this was only done for ungulates, if at all possible; and (4) the time, weather, habitat and place (the term place is used for the geographical location).

The place where the animals had been seen was identified on aerial photographs, pinpointed, and then the serial number of the observation was written on the back. For areas east of the main study area Werdecker's map was also used (Werdecker 1967).

The following environmental factors were recorded: date, time, cloud cover in tenths, wind velocity in five classes according to a simplified Beaufort-Scale (Nievergelt 1966a, p. 11) and direction, exposure towards the wind, whether it was sunny, shady or raining, type of vegetation (range 50 metres around animals), altitude (at place), compass direction which the slope faced (range 10 metres around animals), gradient of the slope (range 10 metres), and the topographical nature of the local area (range 10 metres). When I had recorded all of this data I then concentrated on the feeding behaviour of individuals within that or another recorded group. If Walia ibex or Klipspringer were feeding in an area which was accessible with respect to time I would determine on which plant species the animals were feeding and which plants were avoided (see Chap. 7.5, p. 65).

Priority was given to observations on the Walia ibex at all times, and consequently on a few occasions herds of Gelada baboons and groups of Klipspringer had to be ignored. In such situations where particular groups were visible for a long time or reappeared I usually recorded them several times. For these cases determined each individual or group was not considered more than twice an hour nor more than three times a day in the analyses of behavioural and ecological data. (For the question of limited independency of data see below, Chap. 8.1, p. 69).

Fig. 23. The location of the observation points and their lookouts within the main study area (see Tables 5–7). The various lookouts referring to an observation point are drawn with the *same symbol*. The coordinates correspond to those used in the map of Stähli and Zurbuchen (1978)

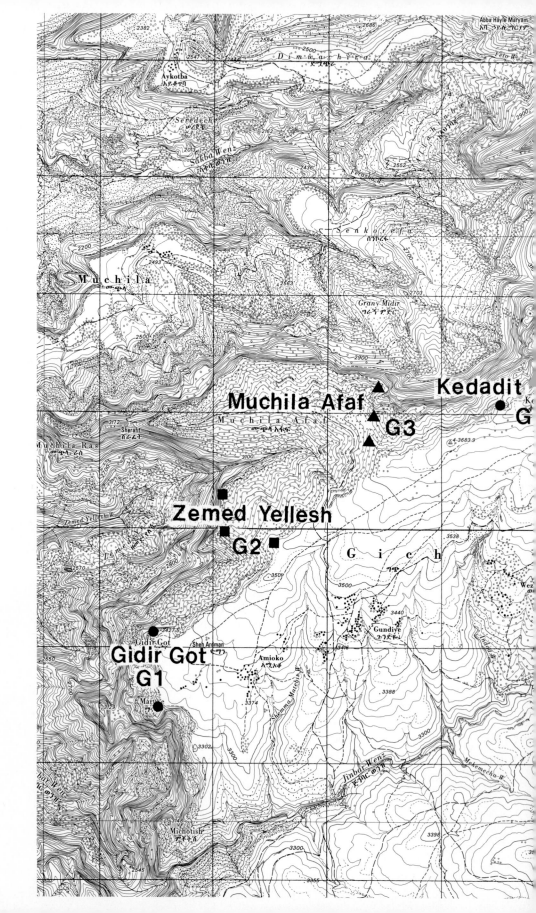

Muchila Afaf

Kedadit Ke

G3

G

Zemed Yellesh

G2

Gidir Got

G1

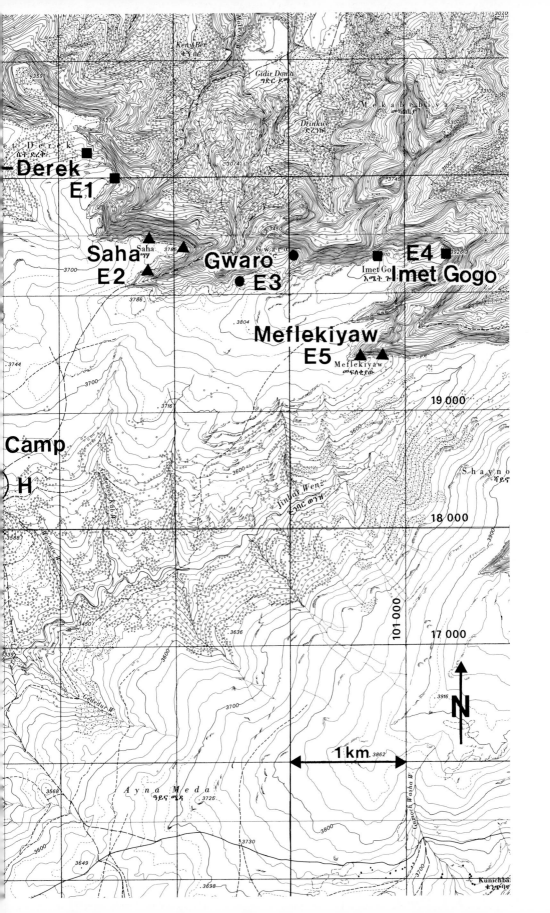

Table 5. The number of visits made to each observation point for each month and for the whole year. Visits of less than 30 min, as well as those when there was heavy fog, were not considered in this table

Observation point		Jan	Feb	Mar	Apr	May	June	July	Aug	Sep	Oct	Nov	Dec	Year
West	Sankaber	1		5										6
Main study area	Gidir Got G1		1	1	1					1		1	1	6
	Zemed Yellesh G2	1	1	1	1	1	1	1		1	1	1		10
	Muchila Afaf G3	4	5	4	3	4	2	2	3	6	6	1	4	44
	Kedadit G4	3	8	3	1	5		2	1	4	4	1	4	36
	Set Derek E1		1	1		1	1	1	1	1	1	1	1	10
	Saha E2	4	5	3	3	5	2	5	1	3	3		5	39
	Gwaro E3					2				2	6	4	2	16
	Imet Gogo E4	2		5				1	1	3	4		3	19
	Meflekiyaw E5	1	2		2			1	1	2	3	2	2	16
East	Amba Ras/Chennek Dirni	1									1	2	2	6
	Silki											3	1	4
All of the OP's		17	23	23	11	18	6	13	8	23	29	16	25	212

Table 6. Numbers of animals of each species recorded from the selected observation points. × means animals were seen, but not recorded.

Observation point		Walia ibex Capra i. walie	Klip-springer Oreo-tragus oreo-tragus	Gelada baboon Thero-pithecus gelada	Gureza Colobus abys-sinicus	Bush-buck Trage-laphus scriptus	Simen fox Simenia simensis	Golden jackal Canis aureus	Serval Leptail-urus serval	Leopard Panthera pardus	Bushpig Pota-mocho-erus porcus	Duiker Sylvicapra grimmia
West	Various OP's Sankaber-Wolkefit	19	8	610	30	1	1	–	–	–	–	1
Main study area	Gidir Got G1	41	12	345	–	–	–	–	–	–	–	–
	Zemed Yellesh G2	120	9	61	7	–	–	1	–	–	–	–
	Muchila Afaf G3	377	162	570	1	1	–	–	–	–	–	–
	Kedadit G4	191	95	529	2	6	3	6	–	–	1	1
	Gich-Camp. House H	–	–	16	–	–	14	19	–	–	–	–
	Set Derek E1	83	28	247	–	–	–	–	–	1	–	–
	Saha E2	713	81	1,357	–	4	1	2	–	–	–	–
	Gwaro E3	130	37	1,062	2	–	1	–	1	–	–	–
	Imet Gogo E4	208	41	570	–	–	1	2	–	–	–	–
	Meflekiyaw E5	127	24	10	–	–	–	–	–	–	–	–
East	Various OP's Amba Ras/Chennek	40	1	×	–	–	–	–	–	–	–	–
	Silki and Ras Dejen-area	25	×	×	–	–	1	–	1	–	–	–
Total number recorded		2,074	498	5,377	42	12	21	30	2	1	1	2

Table 7. Walia ibex observations recorded from the selected observation points. In the columns the following data are given: V stands for the number of visits (see Table 5), TW for the total number of Walias recorded (see Table 6), CW for the codified number of Walias, OW for the total number of Walia observations (groups or lone animals), mg for the mean group size (TW/OW), M for the total number of males and (F) for females including members of various juvenile classes. M>7, >4-7, 3/4, <3 is written for males of the age in years indicated, Y for various undefined young animals, K for kids and F for females. 100·K/F is the number of kids in percentages of the females, 100·K/TW in percentages of the total Walia number recorded

Observation point		V	TW	CW	OW	mg	M	(F)	M >7	M >4-7	M 3/4	M <3	Y	K	F	$\frac{100\ K}{F}$	$\frac{100\ K}{TW}$
West	Sankaber	6	19	17	10	1.9	8	11	4	2	2	0	1	3	4	75.0	15.8
Main study area	Gidir Got	6	41	27	15	2.7	11	30	1	2	8	0	5	4	15	26.7	9.8
	Zemed Yellesh	10	120	85	44	2.7	13	107	1	4	8	5	14	28	48	58.3	23.3
	Muchila Afaf	44	377	271	144	2.6	105	271	7	35	61	28	43	46	116	39.7	12.2
	Kedadit	36	191	148	86	2.2	56	135	2	21	33	11	14	27	70	38.6	14.1
	Set Derek	10	83	63	35	2.4	26	57	11	13	3	4	6	16	27	59.3	19.3
	Saha	39	713	453	207	3.4	169	536	35	78	53	83	69	109	236	46.2	15.3
	Gwaro	16	130	84	45	2.9	42	88	13	18	11	18	3	21	42	50.0	16.2
	Imet Gogo	19	208	123	55	3.8	22	185	4	13	4	12	19	55	93	59.1	26.4
	Meflekiyaw	16	127	74	33	3.8	22	104	5	3	14	13	16	26	47	55.3	20.5
East	Amba Ras/Chennek/ Dirni	6	40	23	11	3.6	18	22	2	5	11	0	3	7	12	58.3	17.5
	Silki	4	25	22	11	2.3	5	20	0	0	4	2	3	4	11	36.4	16.0
All of the OP's		212	2,074	1,390	696	3.0	497	1,566	85	193	212	176	196	346	721	48.0	16.7

7.2 Optimal Time of Day for Spotting Animals and Duration of the Observations

At the start of the fieldwork we had to find the best time for observation, and decide how long these observation periods should be. It soon became clear that the best observation time was during the morning hours. In the evening we were usually less successful, as Table 8 indicates. In this table for Muchila Afaf and Kedadit OP and for two observation period classes, the average number of Walias spotted is given with a standard deviation. For those observations which lasted longer than 90 min, and taking care in the interpretation due to the low number of visits in the afternoon, there is a tendency for greater success in the morning.

Further advantages attached to morning observations were that we could subsequently investigate feeding places (see p. 65), and that weather conditions were fairly good throughout the year – even in the rainy season (see p. 11). For all these reasons the majority of observations were gathered in the mornings.

Table 8. Comparison of the number of Walia ibex spotted at two observation points in the morning (a.m.) and afternoon (p.m.) and for two observation-period classes (30–89 and 90–149 min). $\bar{x}$ stands for the arithmetic mean, n for the number of visits and s for the standard deviation; s was only calculated if n was above 5. * indicates the difference is significant with Wilcoxon rank test (P < 0.05)

OP; a.m. and p.m.		30′–89′			90′–149′		
		$\bar{x}$	n	s	$\bar{x}$	n	s
Muchila Afaf	a.m.	3.83	6	3.43	9.24	17	6.26
	p.m.	3.33	3	–	1.0	3	–
Kedadit	a.m.	0.83	6	0.98	7.77	13	3.54
	p.m.	2.3	10	2.91	3.0*	4	–

Table 9. The average number of spotted Walia ibex at three observation points and various observation periods. At the head of the six data columns the observation period in minutes is indicated. $\bar{x}$ stands for arithmetic mean, n for the number of visits and s for standard deviation; s was only calculated if n was above 5

OP		< 30′	30–59′	60–89′	90–119′	120–149′	≧ 150′
Muchila Afaf	$\bar{x}$	1.75	4.0	3.57	7.75	8.38	7.13
	n	4	2	7	12	8	15
	s	–	–	3.21	5.19	8.48	4.42
Kedadit	$\bar{x}$		1.5	1.9	6.47	8	5
	n		6	10	15	2	3
	s		1.05	3.03	4.09	–	–
Saha	$\bar{x}$	1.2	3.67	5.2	16.38	15.0	15.2
	n	5	3	6	8	12	10
	s	–	–	3.43	11.01	6.67	5.12

In Table 9 the average number of spotted Walia ibex at different observation periods is given. The table shows that observation success apparently increases up to a period of 90 min, but after that the level remains similar even if the spotting time is much further extended. Observation periods of at least 90 min each to around 2 h were attempted at most observation points, as our data soon indicated that the optimal time for observations was roughly at this level. At observation points covering only small geographical units such as Meflekiyaw, the required periods were shorter.

7.3 The Grid Pattern System Within the Main Study Area

A grid reference system was drawn on the topographical map of the Simen Mountains National Park, scale 1 : 25,000 (Stähli and Zurbuchen 1978). The grid divided the area into 10,000 m² fields. Each hectare field was classified according to various ecological parameters with the aid of the map, and terrestrial and aerial photographs. The following information was recorded at the mid point of each field: altitude, gradient and compass direction of slope. For the entire field, values were given based on an arbitrary scale, for various ecological parameters such as the number and quality of ridges, the number of troughs, the number of trees and shrubs and their density, the rockiness of the terrain, the degree of man's influence and the highest possible difference in altitude between this and a neighbouring field. In order to gauge the value of some factors which might have an influence over a range greater than 100 m, e.g., ridges, troughs, degree of forestation and the influence of man, the values for each field and its eight neighbouring fields were added up. I also recorded the number of different compass directions of each field and its neighbours as a further measure of topographical variability. (A list of all these factors and the distinguished classes is given in Table 12, p. 73). In total 5,815 hectare fields were described according to the above values. These descriptions represent a measure of what the Simen mountains offer as a habitat. Relative visibility, given in percentages, was separately recorded for each field overlooked by the three main observation points, Muchila Afaf (G3), Kedadit (G4), and Saha (E2). This represented the probability of seeing a goat-sized ungulate anywhere within the hectare field from a particular observation point. Factors that reduce visibility are long distances, unfavourable direction of the slope, certain topographical features and dense vegetation cover. Relative visibility was estimated using sets of panoramic photographs taken from each observation point as well as aerial photographs. Because the vegetation cover changes little over the seasons in this winterless afroalpine area, the same estimated percentages could be used for the whole year. These estimates of relative visibility were used in order to predict the null-hypothesis in the ecological behaviour of the Walia ibex, the Klipspringer and the Gelada baboon. The distribution pattern of these visibility values was considered to correspond with the pattern of the observed numbers of a randomly distributed animal.

As the location of each group of animals observed was pinpointed exactly on the aerial photographs, the corresponding hectare field on the grid reference system was determined.

7.4 How Comparable and Representative Are Observations Carried Out from the Selected Fixed Observation Points?

Considering human-dependent variables such as spotting time, ability of the observers, and optical equipment and recalling the aforementioned negligible changes in visibility conditions over the year – rainy weather and fog, of course, excluded – the method of observing from fixed observation points implies in fact a high degree of comparability. Within the data of each observation point we can compare relative frequency of animals over time, be it over seasons or over years (see Chap. 9, Estimates of Population Size and Changes of the Walia ibex, and Nievergelt 1971). Comparisons are also possible between various species and classes.

While limiting the number of observation points, the following question becomes especially meaningful. Is the selected area of investigation representative in size and habitat type for the overall Walia ibex range? This question must be answered particularly for the two geographical units covered by the observation points Muchila Afaf, Kedadit and Saha. In fact, as the whole Walia population existing today is restricted to a very limited range within the Simen mountains, some review seems appropriate as to whether this remaining population of Walias lives in an original and typical habitat at all? Is the present distribution not simply the result of a minimal human pressure in this area? Fortunately, due to its ecological heterogeneity the Simen Mountains National Park, although small in size, probably includes all the naturally occurring habitat types which serve to support ibexes between approximately 2,400 and 4,000 m. The Bwahit mountain (4,430 m), in the very east, extends the altitudinal zone open to the Walia ibex up to the highest existing natural regions. Reports from early travellers to Simen document the fact that Walia ibex were present, for instance, on the escarpment around the Gich-plateau when human pressure must have been lower (Maydon 1925; Bailey 1932).

For all these reasons we can assume that the whole region permits the Walia ibex to select its habitat within a drastically limited geographical range, but nevertheless in quite original environmental conditions. With the present population density so low, population pressure which could force animals into a suboptimal habitat can be ignored. Nevertheless it is conceivable that poaching, which did occur more frequently before 1968, has resulted in the Walia frequenting more protected and forested areas (Batcheler 1968).

Within the Simen National Park area during the study period, very roughly 80% of the Walia population existed, but they were ranging in an area of approximately only 50 km². The nine observation points within the main study area overlook about 40% of Walia habitat within the Park boundary and approximately 75% within the main study area between Gich Abyss (Michotish) and Truwata. From most observation points, around 6 km² can be overlooked. With respect to the additional observation points adjoining the main study area, about 75% of Walia habitat within the Park border was observed. The above percentages and approximate sizes are rough estimates and are only used to give an order of magnitude to the reader.

Table 10. The ecological range seen from Muchila Afaf-, Kedadit- and Saha-OP in comparison with the range determined within the whole main study area (see text). The ecological ranges are described by five important environmental factors. Separately for each factor, percentages are given, indicating the relative number of hectare fields the classes represent

Environ-mental factor	Class	OP resp. area				
		Muchila Afaf G3	Kedadit G4	Saha E2	Sum G3+G4+ E2	Whole main study area
		Number of hectare fields of the grid pattern included				
		939	962	970	2,871	5,815
		Number of hectare fields in %				
Altitude, metres above sea level	7 3,700–3,999	1.7	13.4	16.6	10.7	10.2
	6 3,400–3,699	27.2	21.3	14.8	21.0	23.0
	5 3,100–3,399	13.0	5.1	9.7	9.2	13.7
	4 2,800–3,099	14.3	12.5	17.2	14.7	15.4
	3 2,500–2,799	21.9	24.2	33.1	26.5	21.8
	2 2,200–2,499	19.2	21.4	8.4	16.3	13.0
	1 1,900–2,199	2.8	2.1	0.3	1.7	2.9
Gradient of slope	4 >45°	30.8	25.3	41.3	32.5	31.6
	3 30°–45°	30.0	28.0	22.5	26.8	25.4
	2 15°–30°	20.9	18.1	13.1	17.3	21.7
	1 <15°	18.3	28.7	23.1	23.4	21.3
Ridges	0 None	66.7	66.9	60.8	64.8	67.1
	1 Small/few	20.5	20.0	24.2	21.6	23.2
	2 Sharp/many	12.9	13.1	15.0	13.7	9.6
Vegetation	0 Open	25.1	33.7	31.9	30.3	28.8
	1 Single trees/shrubs	38.9	34.5	34.1	35.8	37.8
	2 Open, savanna type forest	27.5	23.9	31.0	27.5	28.2
	3 Forest	8.5	7.9	3.0	6.4	5.2
Influence of man	0 None or almost none	32.2	24.7	23.8	26.9	26.6
	1 Light, occasional	37.4	35.2	38.3	37.0	36.2
	2 Regular feeding pressure	29.9	38.8	29.4	32.7	29.4
	3 Ploughing	0.5	1.3	8.6	3.5	7.9

As data from Muchila Afaf, Kedadit and Saha observation points were specifically used to analyse habitat preferences, it was necessary to check to what extent the overlooked 10,000 m² fields could be considered a typical sample of the whole study area.

In Table 10 in the first column, five important variables that determine the habitat are listed. For each of the three observation points, the table shows the sum of the three and the whole main study area, the relative amount of hectare fields in percentages appearing in each class of the variable. Comparing the values, it can be concluded that the ecological range visible from the three observation points is similar to that of the whole main study area. The values for other parameters not shown here deviate similarly in analogous comparisons.

7.5 Techniques Applied to Investigate Feeding Behaviour of Walia Ibex, Klipspringer and Gelada Baboon

Having once recorded all the groups spotted from an OP, whenever we saw Walia ibexes or Klipspringers feeding at places which we could reach within reasonable time, we concentrated on the feeding behaviour of this particular group or lone animal. Using binoculars or telescope, we noted the number of bites taken at each plant that we could recognize at the distance. Such plants were mainly trees or shrubs. Small species growing within the grass-layer were only rarely identified. Applying this direct observation technique we watched the feeding behaviour of 108 Walia-associations. At the same time, a sketch of the feeding route was made and the feeding places marked as accurately as possible. Subsequently we went to the place where these animals had been feeding. Thanks to the sketch-plan it was possible to find these places with a high degree of accuracy. There we looked carefully for plants that were freshly eaten. Freshly foraged plants could be distinguished from old ones for approximately two hours. Fresh wounds are green and often wet, while old wounds are dry and usually discoloured. At each site we noted the plant species taken, counted the corresponding number of clipped specimens and also recorded those plants which had been ignored by the animals, if they were growing within a 50 cm distance of the feeding places and covering at least 1% of the local area. A total of 50 Walia- and 26 Klipspringer-feeding places were examined (for the technique see also Voser-Huber and Nievergelt 1975).

These techniques have obviously inherent systematic errors and dangers of bias. Quite clearly the number of bites does not properly reflect the amount of plant material taken. Large, thick-leaved plants, as for example *Lobelia rhynchopetalum*, are underestimated, small grasses like *Poa simensis* or *Vulpia bromoides* are overestimated. Therefore the analysis is only semi-quantitative. It is also obvious that a statistical comparison of the results of the two techniques – recording at distance and at the actual feeding place – is not possible. Nevertheless, purely qualitative lists of plant species being foraged by the two ungulates can be amplified with data from both techniques. There is a danger of systematic error because the cut surfaces of different plant species are not equally readily visible. Some comparative tests indicate that searching carefully within a small plot before moving to the next defined plot provided a suitable technique to minimize this error.

For the Gelada baboon this technique was not applicable, although they are also vegetarian (see Crook 1966; Dunbar 1977a; Iwamoto 1979). The later papers have detailed feeding observations for comparison. In contrast to ungulates, Geladas do not clip the plants, but tear them out or off, and sometimes they strip off upper parts such as fruits or dig for bulbs or roots. Thus they do not leave traces which permit easy identification of the selected species. In five situations we watched feeding Gelada baboons at a close distance, noting simultaneously the plants they were selecting.

During the field work most plant species were given a provisional name or number only. This name was changed after having received the list with the proper names. These lists were kindly accomplished in Nairobi and Uppsala on the basis of our herbarium (p. VI). Plants were collected systematically during the preliminary survey as well as during our main field period.

8 Methods for the Analysis of Data on Habitat Selection of the Included Species

8.1 The Variables Used in the Analysis of Data

For each group of animals seen, the habitat in which they were observed was classified according to various environmental factors. These parameters, which referred to each observation and which included data on topography, vegetation and weather conditions, were called observation factors (see Chap. 7.1, p. 55). For geographical location I have used the coordinates of the hectare field in which the group was observed. As each hectare field of the grid-pattern system was classified according to ecological parameters these factors could be used in conjunction with the above observation factors. To distinguish them, they were known as field factors (see Chap. 7.3, p. 62). The following describes which *dependent variables* were used in connection with the observation factors and/or the field factors, the two types of *independent variables*.

For the Walia ibex, the Klipspringer and the Gelada baboon, the features of the selected habitat can be appreciated by a quantitative comparison of the places and habitats where members of the respective species were observed with those areas which were checked for animals without success. The method of using a limited number of fixed observation points in combination with the grid-pattern system covering the study area was the methodical key applied for such a comparison. Separately for each of the three main observation points and for each 10,000 m^2 field a value was calculated based on the total number of animals seen on the field divided by the expected value for random distribution. In this quotient as value for random expectancy the relative visibility for each field was used (see Chap. 7,3, p. 62). Thus, a measure of the relative frequency with which each hectare field had been visited by each of the three species was obtained, to be used as a main dependent variable. The field factors which refer to each hectare field and which lead to a characterization of the environment were consequently the corresponding independent variables. Field factors give an unbiased description of the whole area, and in this field study they have been used to work out the main independent variables influencing habitat selection in the Walia ibex, the Klipspringer and the Gelada baboon. This was done basically by a stepwise multiple regression. This analysis as well as its assumptions, the transformations of data and possible errors are described below under Chapters 8.2 to 8.7. The same independent and dependent variables as in the regression were also used in two-way contingency tables. In such tables the hectare fields are classified in one column by a single field factor – one after the other – and in the other column by the relative frequency of visits of each of the species. All the calculations based on comparisons of the hectare fields are superior because with balanced visibility, an unbiased quantitative comparison of those habitats selected by the animals within the whole investigated area can be attempted or even achieved. But their value is restricted due to the low density of observations because all data per species had to be pooled. Thus the result can only lead to a model of the ecological behaviour of the average Walia, Klipspringer, and Gelada in the yearly average of environmental conditions.

As a further dependent variable, the number of animals observed was used, whether the number in the whole group or the number of animals belonging to each age, sex, and activity class. With these directly observed animal numbers more detailed information could be gained, particularly about habitat selection under varying conditions such as seasonal changes, daytime and weather, and about peculiarities of certain animal classes. It also illuminated the influence of such environmental factors as are decisive according to the multiple regression. In such analyses field factors, as well as observation factors, were used in statistical calculations. Observation factors give a more precise description of the habitat and niche of each species and class, because they are derived from the place where the animals were actually seen, and not just the hectare field in which the animals were standing. A direct comparison is possible with this data, because all the species, classes and conditions concerned were observed using the same techniques. However, it does not give a quantitative description of all the habitats within the study area. This information is given only by field factors.

Consideration was given as to whether a group of animals, regardless of its size or the number of animals in each group, might be used as a variable. Obviously in the first case, where each group, including single animals, was to be counted as one unit, larger groups would be underweighted. In the second case, where each animal counted as a unit, large herds would have an excessive influence on the result. Obviously not all of the group members would independently select this same place where the group was spotted. As Walia ibexes form open groups, changing in size sometimes from hour to hour, the number of "decision units" for the recorded group sizes easily varies between 1 and 5. From the point of view of the pure selection of one particular small place, often a group would have to be considered as one unit. Conversely, in selecting a type of habitat and using a place for feeding or finding space for resting, large associations merit stronger consideration than small groups or lone animals. For the species included in the study a conversion of each recorded group size with reference to the log 2 scale was selected, which seems to accommodate the above aspects in an acceptable way. This conversion was suggested by Preston (1948) in connection with abundance of rare and common species. It resulted in codified numbers as follows:

Observed group size	Codified group size
1	1
2– 3	2
4– 7	3
8– 15	4
16– 31	5
32– 63	6
64–127	7
128–255	8
≥ 256	9

This codified number system was used in the above calculations and gives a measure of the relative frequency of animals spotted on each hectare field, and therein for the Walia ibex, the Klipspringer and the Gelada baboon. In those cases where the number of animals itself was used as a dependent variable, it is stated whether just the number of the groups, the original or the codified number of the animals observed is applied. Real independency of data cannot be achieved if groups of any size in the same way as lone animals, were counted as one unit, because many individuals with possible preferences for certain places are counted several times over the year. This fact may cause a bias to the results. With the low population level of the Walia ibex and the small range of its distribution, such a possible bias cannot be ignored a priori. However, the following findings indicate that no considerable biasing effect need be expected. (1) The seven individually known Walias did not reveal any evidence of specific habitat preferences. (2) As can be seen in Table 32 (p. 151) six of these individuals were spotted from the very regularly frequented Saha- and Kedadit observation points. It is therefore probable that these animals were among the mostly spotted animals. Considering how rarely they were actually recorded (see Table 32) and also in view of the frequencies of different sex-age-classes I estimate that no single animal was spotted so often that it could amount to more than 2% to 4% of all the animals observed. (3) Regressions to analyse the ecological behaviour of Walia ibex as will be shown in Chapter 10 for Muchila Afaf, Kedadit and Saha observation points together, were also carried out separately for these three main observation points, thus for different geographical units, and produced very similar results.

8.2 The Procedure in Applying the Stepwise Multiple Regression

It is assumed that the codified number of animals observed on each field is related to the environmental factors by the following regression model:

$$Y = B_o + B_1 \cdot x_1 \ .... \ B_j \cdot x_j + \ ... \ B_p \cdot x_p + \varepsilon$$

p stands for the number of independent variables.

This relationship is linear in the unknown regression coefficients B_j, but the independent variables x_j can be any, not necessarily linear, function of the environmental factors. The dependent or response variable Y is a function of the codified number of animals. ε is the random error term.

Estimates b_j for the unknown regression coefficients B_j are obtained by minimizing the sum of squares of the error (least-squares method). From the b_j, an estimate for the dependent variable is derived by:

$$\hat{Y} = b_o + b_1 \cdot x_1 + \ .... \ b_p \cdot x_p.$$

This equation can also be used to predict the number of animals on fields for which only the environmental factors have been recorded. For regression theory see for instance Draper and Smith (1966). Calculations were made with BMDP-programmes (Biomedical Programs) on an IBM computer. The best independent variables were selected with a stepwise regression procedure with an F-to-enter value of 1.5 and an F-to-remove of 1.4.

8.3 Assumptions for Regression Analysis

When using the regression procedure, certain assumptions should be checked. Usually they are fulfilled after transformations of the dependent variables. Knowing which assumptions are not exactly fulfilled helps to interpret the result.

a) The data should be representative for the desired purpose. Thus, the prediction equation derived from the observations of the Walia ibex cannot be used for another species or an ecosystem different from the Simen mountains. Neither should it be applied to fields with environmental factors outside the range of those for which the regression equation was derived.

b) The correct model must be chosen. No important variable likely to explain the number of animals observed on a field should be missing. Furthermore, the correct function of the environmental factor should be chosen.

c) The independent variables should be measured with small error as compared with their dispersion. Otherwise the estimated regression coefficients are lower and have greater variance than true ones. This can cause a biologically important, but inadequately quantified variable to be dropped from the regression equation because of a low F-value. Such a variable can be replaced, for instance, by a biologically weak, but properly quantified variable.

d) The error terms ε_i should be independent. This means that observations of one field are allowed to be influenced by observations on neighbouring fields only in a way that can be explained by the environmental factors.

e) The ε_i should be normally distributed, with equal variance σ^2. Under normality the least squares estimates are optimal.

With a number of transformations and controls applied to the data as described in the subsequent Chaps. 8.4 to 8.7, these assumptions may be considered to be sufficiently fulfilled. Some inherent imperfections are discussed in Chap. 8.7.

8.4 Transformation of the Main Dependent Variable

The codified number of animals B observed on different fields is only comparable when all the fields are standardized to the same visibility. B/S can be interpreted as the number of animals present on a field with a visibility of 100%, if they behave randomly with respect to the visible and the hidden part of the grid field.

Any fields with a visibility greater than 20% from one of the observation points Muchila Afaf (G3), Kedadit (G4) or Saha (E2) are included in the analysis. Due to this limitation for fields with relatively high visibility the sample size was reduced to 296 fields. The data set contains 18 fields which have a visibility greater than 20% from the two observation points G3 and G4. Since the part of the field seen from the two observation points is not necessarily the same and the observations from the two points were not made on the same days, they are considered independent and these fields are included twice.

Since B was derived from counting lone animals or groups, it has properties similar to the Poisson distribution. Thus it is assumed that the variance of B is proportional to its expectation [var (B) = C · E (B)]. The expectation is a function of the environmental factors and varies from field to field. On the other hand, the above given assumption (d) implies that for regression purposes the variance of the dependent variable has to be constant. Using the square root transformation stabilizes this variance (Chatterjee and Price 1977).

Let $Y = \sqrt{B/S}$ be the new dependent variable, it can be shown that var

$$(Y) \cong \frac{1}{4 \cdot C} \cdot \frac{1}{S}$$

In order to compensate the dependence of var (Y) on $1/S$, a weighted least-squares regression is necessary, with weights equal to S (see Daniel and Wood 1971). Thus a field whose observations are not so trustworthy because of small visibility has a small weight.

8.5 Relations Between the Environmental Factors

In order to understand and discuss the result of the multiple regression and further statistical correlations, it is important to be aware of relations existing

Table 11. Relationship between the altitude and the following environmental factors; the gradient of the slope (4 classes), the number of shrubs and trees (0: none, open vegetation, 1: single shrubs or trees, 2: open, savanna-type forest, 3: dense forest), and the influence of man and his domestic stock (0: none or almost none, 1: light, occasional, 2: regular feeding pressure, 3: ploughing). For each altitude in 100-metre steps the number of hectare fields is given within the whole study area (n), and separately for each class of the three compared environmental factors. All hectare fields below 2500 m were dropped in the multiple regression analysis (results given in Chap. 10, Tables 15–17)

Altitude in metres above sea-level	n	Gradient of slope				Vegetation: amount of shrubs and trees				Influence of man			
		<15°	15°–30°	30°–45°	>45°	0	1	2	3	0	1	2	3
3,900–4,000	34	26	5	3	–	34	–	–	–	1	30	3	–
3,800–3,899	128	57	49	9	13	121	7	–	–	13	66	49	–
3,700–3,799	431	306	80	17	28	373	46	12	–	21	117	293	–
3,600–3,699	512	253	185	29	45	313	160	37	2	33	266	209	4
3,500–3,599	459	149	200	55	55	162	183	110	4	76	194	139	50
3,400–3,499	369	112	127	46	84	177	106	66	20	94	112	29	134
3,300–3,399	248	46	47	65	90	102	62	35	49	112	49	26	61
3,200–3,299	267	7	46	98	116	43	104	82	38	140	66	52	9
3,100–3,199	280	9	30	95	146	37	148	75	20	141	100	39	–
3,000–3,099	244	4	11	86	143	33	140	59	12	123	100	21	–
2,900–2,999	298	3	21	99	175	40	126	115	17	157	105	35	1
2,800–2,899	351	13	52	109	177	48	136	145	22	135	137	39	40
2,700–2,799	416	32	80	175	129	35	152	196	33	107	161	78	70
2,600–2,699	429	57	94	147	131	38	182	188	21	84	140	143	62
2,500–2,599	424	69	83	143	129	29	209	163	23	75	79	260	10
2,400–2,499	395	76	107	109	103	18	206	158	13	69	81	229	16
2,300–2,399	202	9	16	75	102	23	90	81	8	71	97	34	–
2,200–2,299	159	1	12	53	93	23	66	61	9	61	90	8	–
2,100–2,199	98	–	8	39	51	15	50	33	–	24	68	6	–
2,000–2,099	46	–	3	17	26	6	24	13	3	8	32	6	–
1,900–1,999	25	7	6	10	2	4	2	11	8	–	12	13	–
1,900–4,000	5,815	1,236	1,262	1,479	1,838	1,674	2,199	1,640	302	1,545	2,102	1,711	457

between the fourteen field factors, acting as independent variables in the analysis. In a multiple comparison significant correlations resulted between the gradient, the rockiness of the terrain, the highest possible difference in altitude between this and a neighbouring field, the degree of man's influence and the addition of these degrees of this and the neighbouring fields. These correlations were significant between all the mentioned five variables at a level of $P < 0.01$; in one case only the value just reached 0.01. Correlations were positive between the first three of these variables and the last two, but negative between these two groups. A significant, positive correlation also resulted between the values for the numbers of trees and shrubs and the addition of these values of this and the neighbouring fields ($P < 0.01$). All these relations were to be expected.

In Table 11 the relations of the altitude to the gradient, the numbers of trees and shrubs and the degree of man's influence, which are of a non-linear character are shown. Moreover, the table reflects the features of this type of mountainous area with high plateau, steep rocks and terraces; it indicates the timberline and shows the altitudinal zones of man's activity.

8.6 Transformation of the Independent Variables

The environmental factors are not continuous but nominal (e.g., compass direction of slope) and ordinal (e.g., altitude above sea level) variables. For regression purposes they are separated into "dummy" or "indicator" variables taking only the values 0 or 1. This transformation was chosen because of the existence of nominal variables and also in order to catch non-linear relations between the independent and dependent variables. For example, the variable ridge is given which has the following three possible values: none, few, many. It is separated into two dummy variables (x_1, x_2). They take the following values:

Ridge	x_1	x_2	Values of the prediction equation $\hat{y} = b_0 + b_1 \cdot x_1 + b_2 \cdot x_2$
None	0	0	b_0
Few	1	0	$b_0 + b_1$
Many	0	1	$b_0 + b_2$

This procedure is equivalent to an analysis of variance without interactions. For further discussion of the use of dummy variables see Chatterjee and Price (1977).

In Table 12 all classes or dummy variables into which the environmental factors were divided are listed. The dummy variables are numbered and these numbers are given again in Tables 15 to 17, where the result of the stepwise multiple regression is presented. As can be seen in Table 12 all fields at an altitude lower than 2,500 m above sea level were dropped in the analysis, because no animals were seen in these lowland ranges.

Table 12. List of the environmental factors and the separate dummy variables used in the multiple regression. For each factor one class does not appear as dummy variable. These classes, listed at first for each factor, do not have a variable number. For the Walia ibex (W), as well as for the Klipspringer and the Gelada baboon (K, G) some factors or dummy variables were never significant in various multivariate analyses. These factors, which are indicated with an asterisk in the table, were dropped. This table is also used to define the classes of the environmental factors that refer to each hectare field (field-factors). In addition to the information given in the table for the two factors: quality and number of ridges resp. troughs, both consisting of the classes 0, 1 and 2, I would like to add the following specifications: Class 1: small, soft ridge/trough, but affecting 45° change of compass direction of the slope within field and extending through whole field (or: part of field, but sharp, affecting 90°, or more than 1 small ridge/trough). Class 2: at least one sharp ridge/trough, affecting 90° change of compass direction of slope and extending through whole field (or many small and extended ridges). For specifications on the factors: type of vegetation and influence of man, see also table 11 (p. 71)

Environmental factor; abbreviations are in brackets	No. of variable for		Classes of the factor; the dummy variables
	W	K, G	
Altitude in metres above	–	–	Alt. 2,500–2,799
sea-level (alt.)	1	1	Alt. 2,800–3,099
	2	2	Alt. 3,100–3,399
	3	3	Alt. 3,400–3,699
	4	4	Alt. 3,700–4,000
Compass direction of the slope	–	–	Peak location, not definable
(comp. dir.)	5	5	Comp. dir. N
	6	6	Comp. dir. NE
	7	7	Comp. dir. E
	8	8	Comp. dir. SE
	9	9	Comp. dir. S
	10	10	Comp. dir. SW
	11	11	Comp. dir. W
	12	12	Comp. dir. NW
Gradient of the slope combined	–	–	Grad. $<30°$/diff. alt. ≤100
with difference in altitude to	13	13	Grad. $30°–45°$/diff. alt ≤100
neighbouring field in metres	14	14	Grad. $>45°$/diff. alt. ≤100
(grad./diff. alt.)	15	15	Grad. $<30°$/diff. alt. 200
	16	16	Grad. $30°–45°$/diff. alt. 200
	17	17	Grad. $>45°$/diff. alt. 200
	18	18	Grad. $<30°$/diff. alt. ≥300
	19	19	Grad. $30°–45°$/diff. alt. ≥300
	20	20	Grad. $>45°$/diff. alt. ≥300
Quality and number of ridges	–	–	Ridges 0; none
	21	21	Ridges 1; small, few
	22	22	Ridges 2; sharp, many
Quality and number of troughs	–	–	Troughs 0; none
	23	23	Troughs 1; small, few
	24	24	Troughs 2; sharp, many
Type of vegetation, number of	–	–	Shrubs/trees 0; none
shrubs and trees	25	25	Shrubs/trees 1; single shrubs or trees
(shrubs/trees 0–3)	26	26	Shrubs/trees 2,3; open and dense forest
Amount of rock-coverage	–	–	Rock 0; none
(rock 0–3)	27	27	Rock 1; rock $<50\%$ of area
	28	28	Rock 2,3; rock $\geq50\%$ of area

Table 12 (continued)

Environmental factor; abbreviations are in brackets	No. of variable for		Classes of the factor; the dummy variables
	W	K, G	
Influence of man and domestic stock (man 0–3)	–	–	Man 2,3; considerable, feeding pressure or ploughing
	29	29	Man 1; occasional
	30	30	Man 0; no or almost no influence of man
Sum of class values for ridges in the corresponding field and its neighbours (sum ridges)	–	–	Sum ridges 0
	31	31	Sum ridges 1–3
	32	32	Sum ridges 4–6
	33	33	Sum ridges 7–9
	34	34	Sum ridges ≥ 10
Sum of class values for troughs (sum troughs)	–	–	Sum troughs 0
	*	35	Sum troughs 1–3
	*	36	Sum troughs 4–6
	*	37	Sum troughs 7–9
	*	38	Sum troughs ≥ 10
Sum of class values for numbers of shrubs and trees (sum shrubs/trees)	*	–	Sum shrubs/trees 0
	*	39	Sum shrubs/trees 1–3
	*	40	Sum shrubs/trees 4–6
	*	41	Sum shrubs/trees 7–9
	*	42	Sum shrubs/trees 10–15
	*	43	Sum shrubs/trees ≥ 16
Sum of class values for influence of man (sum man)	–	–	Sum man 0
	35	44	Sum man 1–3
	36	45	Sum man 4–6
	37	46	Sum man 7–9
	38	47	Sum man 10–15
	39	48	Sum man ≥ 16
Number of different compass directions in the nine hectare fields: the corresponding field and its neighbours (n comp. dir.)	–	–	N comp. dir. 1 or 2
	40	49	N comp. dir. 3
	41	50	N comp. dir. 4
	42	51	N comp. dir. 5
	43	52	N comp. dir. ≥ 6
Observation point (OP)	–	–	OP Muchila Afaf
	44	*	OP Kedadit
	45	*	OP Saha
Various reciprocal actions	46–52	*	

8.7 Inherent Imperfections in the Data Being Able to Diminish the Multiple Correlation Coefficient and a Test to Detect Possible but Overlooked Influential Environmental Factors

The transformations of variables described earlier are designed to make the data fulfill the assumptions required for regression analysis. However, several unavoidable and inherent imperfections remain which might increase the error, and they are described below. The resulting mathematical model, given with the regression equation cannot explain the variation in the data completely. The remaining unexplained gap in the result is caused partly by such inherent imperfections. Some of these spring from the fact that the Walia ibex is a rare and endangered species with consequent low observation frequency.

1. Due to topographical heterogeneity with squares of 100 by 100 m a relatively narrow-fielded grid-pattern system was chosen. Due to the low animal density and the resulting low probability of spotting animals at a particular site, I did not observe animals in many fields, even when such fields were perfectly suitable. This was the main reason why fields with a visibility of 20% and less with their low probability for spotting animals were not considered in the multiple regression analysis.

2. None of the animal species included selects its sites to feed or rest only according to environmental conditions. Each animal gets to know its range and tends to frequent familiar places. Thus subsequent observations of the same animals or groups on different days are in fact not entirely independent events; an assumption which was stated in Chapter 8.3 under (d) (p. 70). This phenomenon of individuals persisting in particular places tends to result in higher values of relative frequency for certain fields, as might be expected from the aspect of habitat characterization. Social phenomena like territories have a similar biasing effect.

3. Many of the hectare fields are, in spite of the small size, ecologically heterogeneous, being for instance partly flat and partly steep. A field, which by chance is flat in the center, is not supposed to be attractive to ibexes. Steep rocks in a corner are ignored by classification according to the gradient, and yet might be sufficient to make this corner of the field, and therefore statistically the whole field, attractive to ibexes. Thus heterogeneity within the fields can lead to either higher or lower values than might otherwise be expected. Such errors are therefore responsible for some of the deviations.

4. The regression analysis provides only a model of the ecological behaviour of the average Walia, Klipspringer and Gelada in the yearly average of environmental conditions (Chap. 8.1, p. 68). As will be shown later, ibexes and also to a lesser extent the other two species respond differently to a factor such as the compass direction of the slope in various seasons and times of the day. A similar problem is caused by the phenomenon that habitat preferences in the Gelada baboon change with group size (p. 104). In order to consider such aspects properly, data should in fact be divided according to such variables. Due to the low number of observations this procedure is not possible within this multivariate approach, where a codified number of animals per field is used as a dependent variable. Therefore factors where considerable changes in values occur, depending on conditions such as season, are considered as only fragmentary in a summarized judgement over the year. Fortunately in the afroalpine climate of the Simen mountains there are not such fundamental changes in environmental conditions as in temperate latitudes, and in direct comparisons of data such as contingency tables, considerable seasonal effects are visible in only a few variables.

These aspects undoubtedly lower the multiple regression coefficient and increase the unexplained variation in the data. In the discussion of the unexplained gap, the question must be answered as to whether a particular and perhaps determining environmental factor has been ignored. The probability of such a factor is certainly higher with regard to the Klipspringer and the Gelada baboon than to the Walia ibex, as the selection and classification of environmental factors was done in consideration of the ecology of other Caprini (Buechner 1960; Couturier 1962; Geist 1971a; Nievergelt 1966a, 1968; Pfeffer 1967; Heptner et al. 1966; Schultze-Westrum 1963; Daenzer 1979).

Checking the list of field factors included in the regression (see under Chap. 7.3, p. 62) I am aware of the fact that vegetation is classified in an undifferentiated manner. The completion of the vegetation map by Dr. F. Klötzli will allow further information to be included. However, as the geological situation is relatively homogenous, the main determining factors for the vegetation seem to be included with the selected field factors. This supposition is also supported by preliminary studies on the vegetation in Simen (Klötzli 1975a), and therefore was considered to justify the rough classification of the vegetation. But besides vegetation, there are a number of other possible factors which might have been overlooked or unsuitably classified.

In order to detect any independent variables that had been overlooked, the classified negative or positive residuals for each included hectare field were plotted on the map. In a heterogenous, mountainous area such as Simen we have to expect that some as yet unconsidered environmental factors could be of influence in geographically localized areas, be it a particular vegetation zone, availability of open water, a specific type of human influence or a factor of yet another type. With regard to the plot we can therefore write the following expectation: A clumped pattern of positive or negative residuals indicates the existence of a missing but decisive factor, whereas a random distribution of those residuals makes such a factor unlikely. In the latter instance residuals are caused by random error, by imperfections of data and possibly by ignored factors of little importance. The plot of the residuals is given and discussed after the results of the multiple regression in Chapter 13.

In the next chapter, various approaches to gauge population size of the Walia ibex are presented. As for most parts further below, it is also based on the purely methodical Chapters 7 and 8, but its content is not required to understand the parts following successively. Thus, the reader less interested in this applied and technical problem of population estimates may postpone or drop the lecture of Chapter 9 and proceed to the results of the ecological analysis of the habitats of the Walia ibex, the Klipspringer and the Gelada baboon beginning with Chapter 10, p. 83.

9 Estimates of Population Size and Changes of the Walia Ibex

Methods of assessing population size and changes have to consider the particular environmental situation in Simen, which can be summarized as extremely heterogenous and partly inaccessible but small and overlookable. Among several attempts aimed at gauging the absolute population size, there is none that does not imply a high degree of uncertainty. Three which are believed to have some measure of reliability and to be sufficiently valid to serve as rough estimates are presented in the first part of this chapter. But in fact, in order to plan adequate measures for protection and management it is not necessary to tally the exact number of living animals. Indeed, it is only vital to judge as precisely as possible whether the population remains stable, or whether it is decreasing or increasing. A statistically valid method was proposed in 1971 and subsequently carried out by the Park Warden and the Game Guards in charge (Nievergelt 1971). This method and some results are summarized in the last section of this chapter.

In the first and highly qualitative approach made to get an estimate of the entire Walia population, routine observations carried out from the regularly visited observation points are used. It is based on unusually successful excursions which occurred occasionally as entirely unpredictable events and also on the supposition that considerable population changes between the geographical units do not occur frequently (see p. 138). This supposition is substantiated by the distribution pattern of seven individually known animals (p. 150); in addition there was the impression that I saw, in certain areas subsequently, relatively similar groups. For each season separately I made an estimate for each frequented geographical unit looking at the records of the days with most animals seen. Classes were considered; of especial help were kids apparently persisting for some time in relatively small areas and old males as they could be regularly and clearly recognized. For 1968/69 I estimated for various units from west to east and as an approximate mean over the whole year: Sankaber: 20, Gich Abyss, Gidir Got: 15, Zemed Yellesh and close to Muchila Afaf: 25, Muchila Afaf, Kedadit and west of Set Derek (area covered mainly by OP Muchila Afaf and Kedadit): 30, Set Derek east, Saha, Gwaro, Imet Gogo north: 50 (area covered mainly by OP Saha), Imet Gogo south: 15, Meflekiyaw, Amba Ras, Chennek: 25, outside Park near Silki: 30. According to these figures, the population within the Park area averaged roughly 180 or about 150 to 200 animals. As will be shown later, these estimates were probably too low.

An attempt at a total count of Walia ibex was undertaken by J.P. Müller during his Park wardenship lasting from 1971 till 1973. The occasion arose when the John Hunt Exploration Group under their leader Lt. Col. Tony Streather made an

Endeavour Training Expedition to the High Simens and offered to do a useful task for the National Park (Veitch, R.S. ed. 1972). On February 4 to 5th 1972 J.P. Müller, with 26 people involved, organized a simultaneous count of Walia ibex from 10 different observation points covering the whole escarpment range within the National Park border. The evening count estimated 100 animals observed, the next morning the count was 110. In 1972, 110 animals therefore represented a valid minimum figure. As it is most unlikely that in such a count all Walias living within the Park could have been spotted, the true number was certainly higher (see Müller 1972).

The third method is based on regular observations, as well as on the relative visibility being determined separately for the three main OP's Muchila Afaf, Kedadit and Saha and from those for each hectare field within the coresponding geographical unit. Estimates therefore were made for those areas covered by the three OP's only.

It was postulated that the relation of the mean number of observed animals per OP and excursion (mW_{OPi}) to the total, but unknown, number within the covered geographical unit TW_{OPi} corresponds to the quotient: Sum of the visibilities being estimated for each field over the whole geographical unit (S vis x_{OPi}) divided by the corresponding sum, when each field would be seen with a full visibility of 100% (S vis 100_{OPi}). Suppressing the index OPi (calculations have to be treated for each OP separately, therefore the index OPi was added above for clarity) we can write:

$$mW : TW \approx S \text{ vis } x : S \text{ vis } 100$$

The only unknown factor is the total number of Walia ibexes in the area (TW). Thus we can calculate:

$$TW \approx mw \cdot \frac{S \text{ vis } 100}{S \text{ vis } x}$$

Still, this formula needs some corrections, which are described in the following. The results with three calculated estimates for each of the three OP's are presented in Table 13. It must be mentioned that the two OP's Muchila Afaf (G3) and Kedadit (G4) cover mainly the same area, whereas Saha (E2) is related to the geographical unit adjoining to the east. We have therefore to expect the calculations for G3 and G4 to average relatively similar values. According to the main idea leading to the approach the mean number of Walias spotted (mW) is designed to be a mean number of samples taken at any particular time. In Table 9 (p. 61) it was shown that the number of ibexes recorded at an OP increases steadily up to at least the 90 min usually spent spotting. During this time the area covered was checked carefully two to three times. The steady increase of the total number being seen in the spotting time indicates the fact that animals become visible during the spotting period by moving to a different place; at the same time others pass out of sight but remain counted. Needless to say that animals were only considered as new, if, due to their sex-age classes or their place, a double counting could be excluded. I estimate that the mean number of Walias being observed per visit at an OP is at least twice as high as the theoretical value of all the animals visible at any one moment. Therefore the mean number of Walias spotted was divided by two (see Table 13). It is also presupposed that the places selected by the animals are neither especially open and clearly visible nor well

Table 13. Estimation of the Walia population using the mean of the observed number per OP (mW) and the relative visibility of the respective area (description of the approach in text). * indicates all hectare fields are included with a visibility of $\geq 1\%$ from the corresponding OP. ** as additional restriction only fields above 2,400 m and with no or only sporadic human influence (classes 0 and 1) are included. *** as a further restriction only steep-sloping fields are included; classes $30°-45°$ and $>45°$. c is a correction factor needed in case ***, due to eliminated ibex habitat on less steep slopes

		mW	S vis 100	S vis ×	c	$TW = mW \cdot {}^1/_2 \cdot c \dfrac{S\ vis\ 100}{s\ vis\ ×}$
Muchila Afaf (G3)	*	6.91	93,900	8,159	1	40
	**		58,600	6,644	1	30
	***		41,100	5,542	1.03	26
Kedadit (G4)	*	4.33	96,200	4,718	1	44
	**		49,900	3,106	1	35
	***		33,000	2,485	1.01	29
Saha (E2)	*	12.95	97,000	10,852	1	58
	**		52,200	6,795	1	50
	***		42,300	4,483	1.12	68

covered. I considered ibexes feeding and/or walking – the main basic activities in the morning and evening hours when we were spotting – to cause presumably little bias of this kind. I do not see any evidence that in ordinary feeding or moving, Walias were deliberately hiding or presenting themselves.

As within a geographical unit not all places are suitable for Walias, the quotient S vis 100/ S vis x was changed in two ways in so far as potential ibex habitat was included exclusively. First (**) those hectares below 2,500 m and with heavier human influence (classes 2 and 3) were excluded. No Walia ibexes were observed on fields that were eliminated by this approach. Secondly and additionally (***): only hectare fields at least $30°$ steep remained. In this last calculation a correction factor had to be used, taking into account the number of ibexes observed on less steep slopes (see Table 13).

The total estimates (TW) given in the last column of Table 13 do not claim a high reliability. It is, for instance, quite possible, that, instead of 1/2, in fact 2/3 was a better estimate to include in the equation, changes, which would considerably influence the values in the last column. However, changes can be applied by the reader as the figures leading to the estimates are given in the table. As with this third approach an estimate was made for the areas covered by the three OP's Muchila Afaf, Kedadit and Saha only, an – of course rather speculative – extrapolation for the whole area of the Simen National Park would result in roughly 250 animals.

As was introduced above, none of the three approaches is appropriate to really reach a satisfactory level; too much remains uncertain or arbitrary. However, I felt it is still more valuable to give a vague answer to the question about the population size than none. Fortunately, instead of gauging the actual numbers of Walias it is definitely easier to detect population changes, and such information is of high priority to the conservation authorities (p. 177).

Table 14. Summarized result of the regular Walia counts, designed to detect changes in the Walia population (see text). The arithmetic mean of the counted Walias ($\bar{x}$) and the number of excursions carried out (n) are given for four subsequent periods covering 1968 until 1977 and for the observation points located at the edge of the plateau in the sequence from the western to the eastern border of the National Park. Using the original numbers of Walias counted, χ^2-heterogeneity tests were applied. In the last column, apart from remarks, the level of significance is given only

Location in Simen Park	Observation point (OP)		1968/69 B.N.	1971/73 JP.M.	1973/75 P.S.	1975/77 H.H.	Level of significance according χ^2-heterogeneity test of whole samples and remarks
Western border	Sankaber (OP)	$\bar{x}$	2.8	5.3	4.8	4.5	n.s. due to few regular observations until 1973; but reliable indications for considerable increase after 1969
		n	6	3	446	381	
Main study area	Gidir Got G1	$\bar{x}$	5.0	9.7	4.3	4.1	n.s.
				7.5			
		n	6	7	52	97	
	Zemed Yellesh G2	$\bar{x}$	10.5	15.5	2.9	3.0	P<0.01 decrease and probable stabilisation
				11.3			
		n	10	2	57	77	
	Muchila Afaf G3	$\bar{x}$	6.9	5.8	2.3	2.5	P<0.01 decrease and probable stabilisation
		n	44	8	55	68	
	Kedadit G4	$\bar{x}$	4.3	4.1	2.9	6.8!	P<0.01 decrease, but increase after 1975
		n	36	15	39	45	
	Set Derek E1	$\bar{x}$	7.2	7.4	3.9	4.8	(P<0.1) n.s. probable decrease and recovery after 1975
				7.3			
		n	10	5	35	28	
	Saha E2	$\bar{x}$	12.9	9.7	7.4	7.3	P<0.01 decrease
		n	39	20	53	61	
	Gwaro E3	$\bar{x}$	6.0	8.7	5.5	4.4	n.s. decrease (?)
				6.7			
		n	16	6	37	37	
	Imet Gogo E4	$\bar{x}$	8.8	7.0	8.8	12.2	n.s. tendency to increase (?)
				8.5			
		n	19	5	46	33	
	Meflekiyaw E5	$\bar{x}$	5.8	9.6	5.9	8.5	P<0.05 presumed increase after 1975
		n	16	9	34	38	
Eastern border	Amba Ras/ Chennek (4 OP's)	$\bar{x}$	9.7		10.8	10.0	No regular data before 1973, but presumed increase after 1969
		n	3		398	299	

The theme of the method is the supposition, that if a fixed observation point is used together with similar spotting times, optical equipment and observers, with an increasing population, in general more animals will be seen and vice versa. Between Sankaber and Chennek 15 OP's were selected, including those used during my own field work. All of them were visited several times throughout the year. Thus it was possible – separately for each OP – to compare the numbers of animals actually observed and not an estimated figure. Of course, these regular counts do not give any information about actual population size, but indicate reliably changes in each covered area. Table 14 summarizes the results of the counts for 11 OP's and 4 periods up to 1977. Details and a discussion on possible errors will be given in another publication. The regular counts were carried out for 1968/69 by myself usually accompanied by my wife and/or Ato Berhanu Asfaw and usually one Game Guard, for 1971–73 by J.P. Müller together with Game Guards and for 1973–75 and 1975–77 by groups of Game Guards, but closely supervised by the Park Wardens of the corresponding period: P. Stähli and H. Hurni. As from 1971 Ato Berhanu Asfaw acted as Assistant Warden in Sankaber and was responsible for the counts in that area. In fact, there is a considerable heterogeneity in the observers that requires careful attention; fortunately several trained persons such as Ato Berhanu, Ato Hussein Hassan and Ato Ambaw were involved in all distinguished periods. As a statistical test Chi-square was used on the basis of the result for each single excursion. The level of significance is given in the last column of Table 14. In the table the arithmetical mean per OP and counting period is given for easy comparison of the data. As a general pattern up to 1975 there was an apparent decrease in the central area of the Park but with evidence for increase in the eastern and western marginal zones of the National Park area. For the Sankaber area this evidence is backed by an apparent difference in the chance to spot Walias. In 1968, Walia observations were considered a lucky event, as from 1971 it was almost a question of routine for the tourists when resting at the entrance of the Park at Sankaber. For the Chennek area, it is the impression of the author while discussing the data with the different Park Wardens that Walias must have been more numerous in recent years. Unfortunately, in Chennek no regular counts were carried out before 1973. Thus, this evidence for increase remains in doubt. After 1975 the steady decrease in the central part seemed to have stopped and the population at least partly recovered. As a conclusion, over the whole area and all counting periods the whole population seems to have maintained or regained the level of 1968.

Various estimates on the total number of Walias made between 1973 and 1977 are of the order of up to and around 300 animals. Considering these last estimates, the results of the regular counts to indicate population changes, and the estimate based on the relative visibility of the overlooked areas, it seems likely that my estimates of 1969 which was 150–200 animals, were very probably too low.

The Niche and Habitat of the Walia Ibex, the Klipspringer and the Gelada Baboon

10 The Habitat Selection of the Walia Ibex, the Klipspringer and the Gelada Baboon According to the Frequencies of Their Visits to the Hectare Fields

In this chapter the results of the analysis of data on habitat selection are presented. In the stepwise multiple regression as well as in pairwise comparisons, a dependent variable has been used that is based on how often the hectare fields, visible from the main observation points Muchila Afaf, Kedadit and Saha, have been frequented by the Walia ibex, the Klipspringer and the Gelada baboon. A detailed discussion, as well as a comparison of the habitats selected most often by various species – on an extensive level the Simen fox, the Golden jackal, the Bushbuck and the Colobus monkey are also included – is given in the chapters following.

For the stepwise multiple regression the independent variables were transformed into dummy variables, as explained on p. 72. This procedure, in which each of the fourteen original field factors was divided into several variables, was appropriate because of peculiarities of the data (see p. 72). But this multiplying of variables and separated treatment of the various classes of each factor led to an equation in which naturally the original factors do not appear as a whole, and which is therefore not directly understandable.

For this reason in a preliminary approach, each of the original field factors was compared separately with a dependent variable, a value measuring the frequency of visits of the three mammal species at each visible hectare field. As in the dependent variable used in the multiple regression, the relative visibility was considered. In contrast to the regression, the values of expectancy which are based on visibility were subtracted from the number of animals per field (not divided by) and all fields with a visibility of above 2% were considered (instead of above 20%). Originally the disposition used in this comparison had been selected also in the multiple regression. However, these were two of the subsequent modifications that have helped to fulfil the assumptions of the regression sufficiently: to divide the observed numbers of animals by the expectancy value and to eliminate hectare fields with a low visibility.

The summarized result of the preliminary comparison for the Walia ibex, the Klipspringer and the Gelada baboon and for ten factors is shown in Fig. 25. In Fig. 24 the principle of all the diagrams is explained. Each ray stands for an environmental factor, and in Fig. 24 the respective portions of the classes correspond to the range of the environment, which would be the null hypothesis. In Fig. 25 those diagrams on the right are based on the fields in which the value –

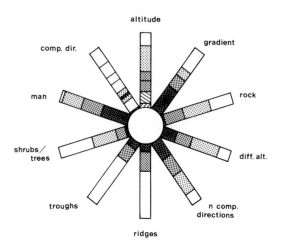

Fig. 24. Ten field factors, each represented by a separate ray, that were used in Fig. 25 for comparison of the relative preferences by the three mammals. In this introductory figure within each factor the respective portions of the classes distinguished correspond to the overall situation in the study area; thus it is based on the frequencies given by the 5,815 hectare fields. Listed in centrifugal direction, in the following the classes of each of the ten factors are given. For a definition of the classes see also Table 12 (p. 73); the abbreviations of the factors used in the figure are printed in italics in the following text. *Gradient* of the slope: $>45°$ (*dark grey*), $30°–45°$, $15°–30°$, $<15°$ (white); amount of *rock*-coverage: 3 (rock only, *dark grey*), 2 ($\geq 50\%$), 1 ($<50\%$), 0 (no rock, *white*); largest *diff*erence in *alt*itude to a neighbouring field in metres: ≥ 300 (*dark grey*), 200, 100, 0 (*white*); number of different *compass directions (ncomp)* in the corresponding field and its neighbours: ≥ 5 (*dark grey*), 4, 3, 1/2 (*white*); quality and number of *ridges*: 2 (sharp, many, *dark grey*), 1 (small, few), 0 (none, *white*); quality and number of *troughs*: 2 (sharp, many, *dark grey*), 1 (small, few), 0 (none, *white*); amount of *shrubs* and *trees*: 3 (dense forest, *dark grey*), 2 (open forest), 1 (single shrubs or trees), 0 (none, *white*); influence of *man* and domestic stock: 0 (none, *dark grey*), 1 (occasional), 2 (regular feeding pressure), 3 (ploughing, *white*); *comp*ass-di*r*ection of the slope: peak location (*dark grey*), N (*white*), NE (*grey*), E (*dark grey*), SE (*grey*), S (*white*), SW (*white*), W (*white*), NW (*white*); *altitude* in metres above sea-level: 1,900–2,499 (*hatched*), 2,500–2,799 (*hatched*), 2,800–3,099 (*dark grey*), 3,100–3,399 (*dark grey*), 3,400–3,699 (*grey*), 3,700–4,000 (*white*)

number of animals minus expectancy – is above 0.5. These are therefore the fields which were frequented by the animals. In those diagrams on the left the corresponding value was below -0.5. Thus, they represent fields in which no animals were seen or fewer than expected. In a comparison of the diagrams for each species we can see, as summarized on p. 88, some habitat characteristics to be favoured ($+$) or avoided ($-$) by the animals (for the classes of the environmental factors the same abbreviations are used as in the dummy variables, see Chap. 8.6 and therein Table 12, p. 73).

These tendencies summarized, based on Fig. 25, give a general impression of habitat preferences in the three species. It is important to be aware of the fact that these results are derived from single comparisons only. Some of the visible effects may therefore have been caused by relations between environmental factors. Special care is necessary because of the correlation described above between the

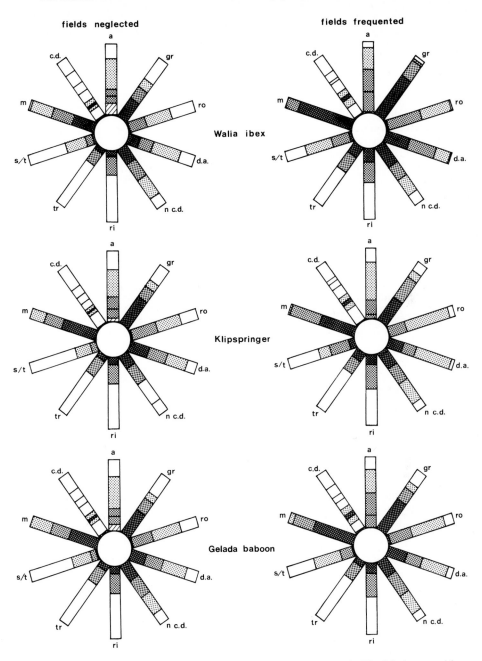

Fig. 25. The diagrams presented are of the same type as the diagram in Fig. 24, thus consider the same ten field factors; in contrast they are based on the hectare fields visible from the observation points Muchila Afaf, Kedadit and Saha only. For the Walia ibex, the Klipspringer and the Gelada baboon the diagrams on the *right side* show the frequency pattern for those fields, where the number of observed animals minus expectancy is above 0.5 (fields frequented), the diagrams on the *left side* represent those fields where the corresponding value is below −0.5 (fields neglected)

gradient, the rockiness of the terrain, the difference in altitude to a neighbouring field and the degree of man's influence (see Chap. 8.5, p. 71).

For the Walia ibex

Altitude	(+) 2,800–3,390;
	(−) below 2,800, as well as above 3,700
Gradient of slope	(−) ≤30°; (+) >45°
Rockiness of terrain	(−) 0 – no rock; (+) 2, 3
Difference in altitude to neighbouring field	(−) ≤100 m
Number and quality of ridges	(+) 2-sharp, many
Number and quality of troughs	(−) 0 – no troughs; (+) 2-sharp, many
Influence of man and/or domestic stock	(−) 2, 3 – grazing or ploughing
Compass direction of the slope	(+) generally towards the east,
	(−) towards the west

For the Klipspringer

Altitude	(+) 3,100–3,699; (−) below
Gradient of slope	(+) 30–45°
Rockiness of terrain	(−) 0; (+) 1
Numbers of shrubs and trees	(+) 2, 3 – light savanna-type forest and forest
Influence of man	(−) 2, 3

For the Gelada baboon

Altitude	(+) 2,800–3,399 m
Gradient of slope	(+) >45°
Rockiness of terrain	(−) 0; (+) 1, 2, 3
Difference in altitude to neighbouring field	(−) 0; (+) ≥300
Number and quality of troughs	(+) 2
Influence of man	(−) 2, 3
Compass direction of the slope	(+) generally towards the east

As stated earlier, the main approach to discover the governing factors in habitat selection of the Walia ibex, the Klipspringer and the Gelada baboon was carried out by a stepwise multiple regression (see Chap. 8). In Tables 15, 16 and 17 the results are presented for the three mammals. Each table lists the multiple correlation coefficient, the result of the analysis of variance with the sum of squares, the degree of freedom, the mean square and the F-value, and all the independent variables considered in the final equation with their coefficient, the standard error of the coefficient and the F-value achieved. The multiple correlation coefficient R of 0.62 for the Walia ibex, 0.52 for the Klipspringer and 0.51 for the Gelada baboon, indicating the amount of deviation explained by the significant and listed independent variables, seems reasonable if one considers the various inherent imperfections in the data (see Chap. 8.7, p. 74). In the following, the main results of the multiple regression are described for each species separately.

The Walia ibex, as shown in Table 15 on the average frequents mainly those altitudes from 2,800 to 3,400 m where the slopes face towards the east, but less those facing west and southwest (negative coefficients), and with strong evidence slopes with a gradient of 45° and more and – with relatively weak evidence (F 1.94) – slopes with a low gradient if the difference in altitude to a neighbouring field is great. This situation is characteristic for fields on the high plateau situated at the very edge of the escarpment. The Walia ibex seems to prefer fields with troughs,

Table 15. The result of the stepwise multiple regression for the Walia ibex carried out with dummy variables. The abbreviations of the variables used in the tables are listed and explained in Table 12. Only the variables that were selected in the stepwise regression procedure are shown in the table. The given sequence does not correspond to the order of selection

| Multiple correlation coefficient | | $R = 0.621$ | |
| coefficient of determination | | $R^2 = 0.385$ | |

Analysis of variance table (ANOVA table)

	Sum of squares	df	Mean square	Total F
Regression	244.842	22	11.129	7.77
Residual	390.985	273	1.432	

Variables in equation

Number of variable	Variables	Coefficient	F
	Y – intercept	−0.547	
1	alt. 2,800–3,099	0.516	2.34
2	alt. 3,100–3,399	0.888	11.29
7	comp. dir. E	0.960	6.28
10	comp. dir. SW	−0.571	3.84
11	comp. dir. W	−0.366	2.90
14	grad. >45°/diff. alt. ≤100	0.520	3.09
17	grad. >45°/diff. alt. 200	0.834	9.96
18	grad. <30°/diff. alt. ≥300	0.647	1.94
20	grad. >45°/diff. alt. ≥300	0.557	4.03
21	ridges 1	−0.208	1.45
22	ridges 2	−0.495	3.30
23	troughs 1	0.260	2.04
24	troughs 2	0.716	6.18
26	shrubs/trees 2, 3	−0.478	2.59
27	rock 1	0.427	6.04
30	man 0	0.613	6.11
31	sum ridges 1–3	0.339	3.94
33	sum ridges 7–9	0.501	4.72
36	sum man 4–6	0.574	5.85
37	sum man 7–9	0.813	11.88
40	n comp. dir. 3	−0.239	2.21
45	OP Saha	0.640	12.90

but not ridges – in simple regressions there is a positive, although weak, effect for ridges too – it favours terrain which is slightly rocky and where the influence of man or of his domestic stock is non-existent or slight. Open and more dense forests seem to be frequented rarely but a strong positive contribution is indicated for the area visible from Saha observation point and for regions where man's influence is slight and irregular (addition of the degree of man's influence in a field and its neighbours 7 to 9). An analysis of the particular fields which are responsible for this unexpected effect has shown that they are concentrated to a great extent in an easterly exposed trough where the footpath from Saha to Dirni, which was used infrequently, goes through, and where ibexes were seen quite often. Due to this footpath, the degree of man's influence in many of these fields is 1, thus for sets of nine fields the class 7 to 9 is well represented.

Table 16. The result of the stepwise multiple regression for the Klipspringer carried out with dummy variables (see also Table 15)

Multiple correlation coefficient			R $=0.518$	
Coefficient of determination			$R^2=0.268$	

Analysis of variance table (ANOVA table)

	Sum of squares	df	Mean square	Total F
Regression	62.104	14	4.436	7.34
Residual	169.806	281	0.604	

Variables in equation

Number of variable	Variables	Coefficient	F
	Y – intercept	-0.259	
3	alt. 3,400–3,699	0.482	13.57
4	alt. 3,700–4,000	0.345	5.37
6	comp. dir. NE	-0.324	2.18
10	comp. dir. SW	-0.392	4.67
11	comp. dir. W	-0.269	3.66
14	grad. $>45°$/diff. alt. ≤ 100	0.635	11.67
18	grad. $<30°$/diff. alt. ≥ 300	0.612	4.51
23	troughs	-0.302	6.97
26	shrubs/trees 2,3	0.578	9.79
27	rock 1	0.377	7.88
28	rock 2,3	0.363	5.98
46	sum man 7–9	0.232	3.16
49	n comp. dir. 3	0.165	2.27
51	n comp. dir. 5	0.316	6.28

The result for the Klipspringer is given in Table 16. The most frequented altitude is between 3,400 and 4,000 m; steep slopes with a small difference in altitude to a neighbouring field and slopes with a low gradient but with a great difference in altitude to a neighbouring field seem to be preferred, as well as open to dense forests and rocky terrain. The topographical relief seems to be of importance as all classes of the factor, the number of different compass directions in the corresponding field and its neighbours, with the value 3 and above, have a positive coefficient; the classes 3 and 5 (variables 49 and 51) have entered into the equation and their contribution is therefore significant.

Table 17 lists the result for the Gelada baboon. Some of the significant variables are evident and plausible, such as positive coefficients to generally east-facing slopes, absence of or rare influence of man, zones with many different compass directions of the slopes, and the negative coefficient for moderate gradient and forested areas. Some further effects are difficult to understand, mainly the positive and negative coefficients in slopes of steep gradient, and the positive contribution for generally forested areas (variable 43) in combination with the negative coefficient of variable 26: open to dense forests in the hectare field. It is possible that such effects are caused, for instance by the reversing habitat selection of groups of different sizes (see p. 102), a fact which seems most likely to be responsible for the generally weak level of significance.

Table 17. The result of the stepwise multiple regression for the Gelada baboon carried out with dummy variables (see also Table 15)

Multiple correlation coefficient			R =0.510	
Coefficient of determination			$R^2 = 0.260$	

Analysis of variance table (ANOVA table)

	Sum of squares	df	Mean square	Total F
Regression	102.626	16	6.414	6.13
Residual	291.811	279	1.046	

Variables in equation

Number of variable	Variables	Coefficient	F
	Y–intercept	0.108	
5	comp. dir. N	−0.347	5.30
6	comp. dir. NE	0.457	2.49
7	comp. dir. E	0.498	2.59
8	comp. dir. SE	0.599	4.09
13	grad. 30–45°/diff. alt. ≤100	−0.514	4.91
14	grad. >45°/diff. alt. ≤100	0.644	6.72
17	grad. >45°/diff. alt. 200	−0.469	5.75
19	grad. 30–45°/diff. alt. ≥300	−0.851	7.70
26	shrubs/trees 2,3	−1.310	13.69
28	rock 2,3	−0.248	2.18
29	man 1	0.652	13.98
30	man 0	0.739	11.13
38	sum troughs ≥10	1.132	10.84
43	sum shrubs/trees ≥16	1.225	11.59
44	sum man 1–3	−0.379	3.78
47	sum man 10–15	0.233	1.92

Recalling Tables 15–17 and the corresponding description above, it is essential to accept the presented results carefully. As can be demonstrated by deliberately eliminating certain variables in the regression, further changes in the resulting equation occur. These changes affect particularly variables of weak evidence, but the major variables are also modified in their weight: e.g., by eliminating the variables 5–12 in the regression for the Walia ibex that stand for the various compass directions (Table 12), as compared to the equation in Table 15, five further variables are dropped (F values 1.45–3.30) and three new variables are selected (F values 2.13–4.03). Thus, care is recommended mainly while discussing variables of a low F-value. Additionally, it is useful to compare the data in the equation with the diagrams presented above (Fig. 25) and with further results given below.

In the following chapter the main independent factors are investigated separately with results based on observation factors, as well as on the direct and codified number of observed animals (see Chap. 8.1, p. 68). These data will be presented and discussed with reference to the result of the stepwise multiple regression. In connection with the independent factors, as well as in particular subdivisions, further influential factors are examined such as seasonal effects, daily effects, weather, group size and activity.

11 The Habitat Selection of the Walia Ibex, the Klipspringer and the Gelada Baboon with Regard to Particular Environmental Factors

11.1 The Gradient of the Slope

The gradient of the slope is correlated with a number of other environmental factors (see Chap. 8.5, p. 71). To one of these, the greatest difference in altitude to a neighbouring field, an interaction became evident. In a cross-table, the frequencies of the Walias observed were compared with the frequencies resulting from the environmental situation in Simen. This comparison is shown in Table 18. Very obviously, steep slopes as well as fields with a high difference in altitude to a neighbouring field are much frequented. It shows further that fields with a low difference in altitude turn out to be positive, if the gradient is above 45°, and that fields of a lower gradient have a positive value, if the difference in altitude to a neighbouring field is at least 300 m. For this reason, out of the two factors

Table 18. The (codified) number of Walias (w) observed compared with expectancy values (e), which are representative for the area in Simen, presented in a cross-table, where the gradient and the greatest difference in altitude to a neighbouring field are opposed. The expectancy values were calculated on the basis of the 5,815 hectare fields of the study area. A plus was written if the Walia numbers exceeded expectancy, a minus, if they were smaller

			Gradient of slope				
			<15°	15°–30°	30°–45°	>45°	Sum
Greatest diff. in	≤100 m	w	18−	29−	117−	138+	302−
altitude to		e	272	258	221	131	882
neighbourfield	200 m	w	8−	16−	61−	392+	477+
		e	11	24	90	184	309
	300 m	w	2+	19+	64+	309+	394+
		e	1.9	5	17	74	98
	≥400 m	w	26+	6+	21+	110+	163+
		e	2	4	8	33	47
	Sum	w	54−	70−	263−	949+	1,336
		e	287	291	336	422	1,336

combined dummy variables have been used in the regression. In Klipspringer and Gelada baboon, those interactions were less pronounced.

The gradient was considered twice as an environmental factor. As a field factor it indicates the gradient in the middle of the hectare field where the animals were observed, and as an observation factor it states the gradient at the exact location of the animals. Obviously these two gradients need not necessarily be identical. In order to estimate the decisiveness of the gradient factor for the three species, I checked within each class of the "field gradients", which is the number of the classes according to the observation gradients. The remarkable result of this comparison is given in Table 19. For the Walia, whatever gradient class we examine, according to the field factors most animals select steep slopes as the local habitat. This fact stresses the positive contribution of steep slopes in the multiple regression which is based on field factors only (Table 15) and the strong effect visible in the comparison of Fig. 25. For the Klipspringer the result seems to be very close to the theoretical expectation: the observation factors most often confirm the class of the field factors. For the Gelada baboon in the steeper two classes the observation factors confirm the preference for steep slopes, whereas in the two classes below 30° an

Table 19. Comparison of the gradient value according to the field factor and the observation factor in the Walia ibex, the Klipspringer and the Gelada baboon. (For distinction of the two factors see Chap. 8.1, p. 67). The table lists the number of the observation-gradient classes in percentages separately for each field-gradient class. The highest percentage value for each class and species is given in bold types

Gradient of slope according to			Walia ibex	Klipspringer	Gelada baboon
Field factor	Observation factor				
<15°	Relative amount in %	<15°	26.5	**70.0**	**86.8**
		15°–30°	12.2	10.0	5.9
		30°–45°	18.4	20.0	2.4
		>45°	**42.9**	0	4.9
	n		49	50	204
15°–30°	Relative amount in %	<15°	10.0	19.3	**48.9**
		15°–30°	28.6	**35.5**	11.3
		30°–45°	24.3	33.9	39.8
		>45°	**37.1**	11.3	0
	n		70	62	88
30°–45°	Relative amount in %	<15°	4.6	9.4	14.4
		15°–30°	16.8	27.5	19.2
		30°–45°	37.0	**49.3**	22.1
		>45°	**41.6**	13.8	**44.3**
	n		262	138	104
>45°	Relative amount in %	<15°	0.6	5.9	4.3
		15°–30°	3.3	15.9	9.3
		30°–45°	24.0	**40.6**	32.1
		>45°	**72.1**	37.6	**54.3**
	n		948	170	346

opposite tendency towards plain areas should be mentioned. It is the impression of the observer that Geladas were seen either on the plateau in the vicinity of the safe escarpment or else on very steep slopes. Thus, the findings recorded above coincide with this impression.

Both Tables 18 and 19 indicate the importance of the gradient as a proximate or indirect environmental factor. In Chapter 12 the habitat of seven mammal species is compared on an extensive level. In the cross-tables applied there, the gradient is considered as one variable. Thus, more data on this factor will be found there (Table 25, p. 112), as well as in Chapter 11.5 (Table 24, p. 104) on the relations between habitat and group size, where the same frame of cross-tables is used.

11.2 The Compass Direction of the Slope with Daily and Seasonal Effect

According to the comparison presented in Fig. 25, as well as in the result of the multiple regression (Table 15), the Walia ibex frequents east-facing slopes more often than west-facing slopes. It was at first thought that this result might relate to the fact that the majority of observations were carried out in the morning hours, when east-facing slopes are in sunlight. In the Alpine ibex, a different pattern of frequented compass directions was observed in the morning and in the afternoon, but this pattern was also dependent on the season. In winter there was a preference for sun-facing slopes at both times of the day, but this was not the case in summer, when sun-facing slopes became relatively less attractive (see Nievergelt 1966a, p. 67, Fig. 36). In both the diagrams of Fig. 26 the relative number of Walia ibexes observed for each compass direction is shown for the morning and afternoon hours. In the right diagram the field factors were considered, in the left one, the observation factors. In both diagrams two expectancy patterns for random distribution are drawn, which are based on two different assumptions (see legend to Fig. 26). We can conclude: In the mornings, the east- and southeast-directed slopes were frequented relatively more often, and in the afternoon the west- and south-west-directions were relatively more attractive. But in comparing this changing pattern with the expectancy of random distribution, we have to conclude that in the east directions – particularly south-east – the expectancy values remain generally below the morning and afternoon values, whereas in the west-directed slopes, the expentancy values are generally above. This result indicates that the time of day actually seems to have an effect, but it is not sufficient to explain the result of the multiple regression. What else could have caused this preference for east-facing slopes? It is a preference which can also be seen in the Gelada baboon. The reason may possibly be found in the precipitation. In Fig. 7, p. 14, it was shown that the predominant wind directions in the rainy season are east and northeast, an observation which coincides with the reported findings of H. Hurni, who measured direction and inclination of the rain (p. 14). It follows that in general east-facing slopes receive more precipitation than those directed to the west. This climatic situation supports the impression of the observer, that slopes and particularly troughs which are directed to the east have particularly vigorous, abundant and fresh vegetation with much *Alchemilla rothii*, *Arabis alpina*, *Simenia acaulis* and

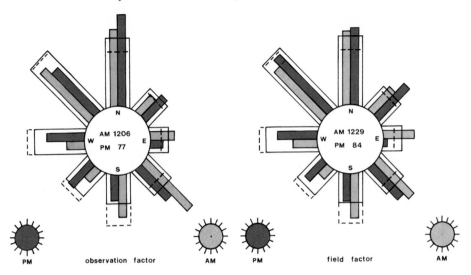

Fig. 26. A comparison of the Walias observed in the morning and in the afternoon according to the compass directions. In the *right diagram* the field factor is used for the compass direction, in the *left* one, the observation factor (explanations see Chap. 8.1, p. 67). The pattern in the morning is shaded more lightly, as is the sun in the morning position (A.M.), the pattern and sun for the afternoon (P.M.) are both darker. Two expectancy patterns for random distribution are drawn. The one with *solid lines* is basing on all hectare fields in Simen with no or occasional influence of man only; the other with *dotted lines* is based only on the fields visible from the three main observation points Muchila Afaf, Kedadit and Saha

Festuca macrophylla, all species which are taken by the Walia ibex. Troughs directed to the west were generally drier, as indicated, for instance, by the more abundant presence of members of the Labiatae such as *Thymus serrulatus, Satureja pseudosimensis* and *Salvia merjamie*; all plants, which were never recorded as eaten by the Walia ibex (see p. 163). Thus, it seems likely that the immediate factor causing the preference of the Walia ibex for east-directed slopes is the vegetation. This interpretation is also supported by the preference of the Walia ibex for troughs, but not for ridges as indicated in Table 15. The vegetation in troughs is definitely better than on the ridges and the Walia ibex obviously selects more large-leaved plants than, for instance, the Klipspringer (see p. 164).

In Fig. 27 the relative numbers of ibexes observed on slopes which are directed southeast, south and southwest are shown separately for each month and for the whole year. The monthly values are given with a confidence limit (P = 0.01). The pattern is compared with the rainy season as well as with the expectancy for random distribution. As in Fig. 26, two estimates for expectancy are given. Obviously, shortly after the rains, the number of ibexes on generally south-facing slopes increases and then drops toward the end of the dry season. The pattern can be confirmed with a test for time series proposed by Dr. E. Batschelet. There is a run of 6 months above the median (September to February) and another below (March to August): the level of significance therefore is P < 0.02 (see Nievergelt 1974, p. 327). The arithmetic mean over the year, as well as the median, is relatively close to

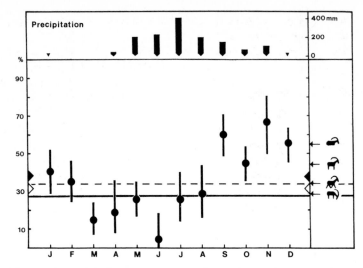

Fig. 27. The numbers of Walias in percentages observed over the year on slopes directed to the southeast, south and southwest including those on peaks. For each monthly amount the confidence limit is drawn (P = 0.01). Two expectancy levels for random distribution are drawn with a *solid* and a *dotted line*. The lower one is based on all hectare fields with no or occasional influence of man only, the upper one is based on the hectare fields visible from the three main observation points (see also Fig. 26). The arithmetic mean (*solid angle symbol*) and the median (*open angle*) for the whole year are given, as well as the yearly mean for lying, standing, walking and feeding animals. The precipitation pattern is based on the figures given in Table 1 (p. 12)

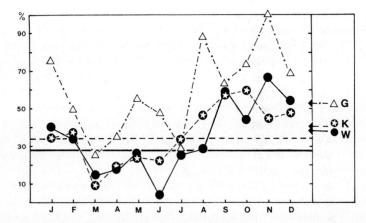

Fig. 28. The numbers of Walias, Klipspringers and Geladas in percentages observed over the year on slopes directed to the southeast, south and southwest including those on peaks. The number for the whole year is indicated on the *right side*. The monthly values were connected for graphical reasons only. The expectancy levels are the same as in Fig. 27

the expectancy level. On the right the yearly mean for animals lying, standing, walking and feeding is given with symbols. Whereas walking animals are close to the random expectancy, ibexes generally seem to prefer drier places for lying and those slopes least exposed to the sun for feeding. This differentiation leads to the most plausible interpretation of the monthly pattern. The vegetation in south-facing slopes dries out faster in the dry season; at this time, the freshest food is generally found in north-directed slopes. South-facing slopes soon become attractive, when, in the declining rainy season, the meadows are greener.

As Fig. 28 shows, the Klipspringer and the Gelada baboon show a similar pattern in the yearly cycle. For the Klipspringer the fluctuations may be slightly smoother than in the Walia ibex, but they are still surprisingly apparent if one considers the territoriality of the species. In the Gelada baboon the generally higher number of south directions must be mentioned.

The changing effect of the compass direction over the year (Figs. 27 and 28) and over the day (Fig. 26) could not be considered in the multiple regression, as described in Chapter 8.7 (p. 75).

11.3 The Altitude and the Vegetation Over the Seasons

The data presented and discussed in this chapter, based on the codified number of directly observed animals, are biased in so far as animals at different altitudinal as well as vegetational belts could not be spotted with the same probability. Basically, animals at lower altitudes, as well as in denser vegetation, were less easily observable. Therefore, and in contrast to the data of the multiple regression, where different visibility was compensated for, lower altitudes and denser vegetation are underweighted. However, comparisons between the three mammal species as well as between the seasons are possible.

In Fig. 29 in the uppermost diagrams the average altitude over the year is shown; the diagrams below treat the vegetation. In the middle ones the presence and absence respectively, and the densities of trees and shrubs are considered; in the lower ones, the grass types frequented are shown. As was seen in the regression analysis, the Walia ibex frequents, on average, lower altitudes than the Gelada baboon and more particularly than the Klipspringer (see Tables 15 and 16). In all three species, but least apparently in the Klipspringer, there is a general tendency to stay at higher altitudes before and at the onset of the rainy season. The rainy season is indicated in the figure (see Table 1, p. 12 for precise data on the rains). In the Gelada baboon, where there is a sudden drop in the rainy season, the curve coincides with the findings of Dunbar (1977a, 1978a) who reports that Geladas feed on the freshly grown grass shoots in the escarpment after the rains, but feed predominantly on rhizomes and roots at the end of the dry season. This is a behaviour that I observed regularly in that season on the high plateau, and which corresponds therefore to the high altitude indicated in Fig. 29. However, in all three species the range in altitude utilized is wide, and in contrast to the averages given, does not seem to differ fundamentally over seasons.

Concerning the vegetation types and comparing the species, we can see that the Walia ibex rarely frequents open areas with Giant Lobelias and that the Gelada baboon seems to avoid forested zones more than the two ungulates. Long grass vegetation which consists mainly of *Festuca macrophylla* is represented most in the diagram for the Walia ibex and least in that for the Gelada baboon. The

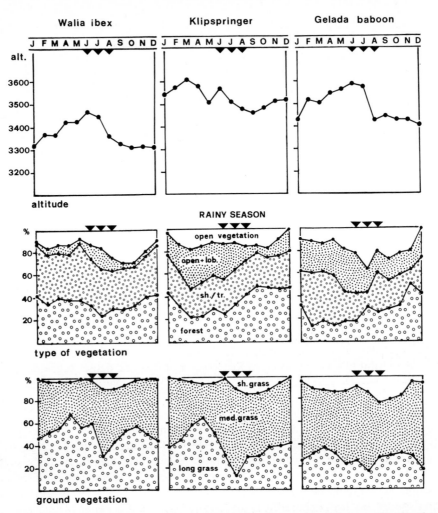

Fig. 29. The average altitude and the amounts of various vegetation types for the three mammal species over the season. In the three upper curves the average altitude is given. In the diagrams in the *middle* the average and relative amounts of four different vegetation types are drawn: open, open with Giant Lobelias (*open + Lob*), single shrubs or trees (*sh./tr.*), and open savanna and forest (*forest*); see also Table 24 (p. 104). In the *lower* diagrams the average amounts of different ground vegetation are shown. Long-grass vegetation, medium- and mixed-grass vegetation and short-grass vegetation are distinguished (see for further information Fig. 31, p. 113. The peak of the rainy season is indicated in all diagrams. In order to smooth the shape of the curves, in the whole figure, for the monthly averages, a gliding arithmetic mean was used. The average given for each month is actually the average of that month and the two months on either side of it

combination of the factors *Festuca macrophylla*-long-grass vegetation with steep slopes of a gradient above 45° is fairly common in troughs; thus a further factor is given which has a positive regression coefficient in the multiple regression. In the yearly cycle it is particularly evident that in all three mammal species the amount of short-grass vegetation used, which consists of species such as *Festuca abyssinica*, *Danthonia subulata*, *Poa simensis* and *Aira caryophyllea*, is highest in and after the rains. These vegetation types, usually growing on little soil only, dry out relatively soon after the rains and obviously lose their attraction in the course of the dry season.

Further data on the vegetation types preferred by each of the species are given in Chapter 12. These data are based on the yearly averages only.

11.4 The Relief, the Weather and the Activity

The relief or the topograhpical nature of a place may have a direct influence on the habitat selection of the animals with respect to thermal comfort, the possibility of being able to overlook the terrain or of being protected, and it may also have an indirect influence through the vegetation. As will be seen in the Walia ibex, in which I considered the activity of the observed animals – this was not done in the other two species – there is most clearly a direct and an indirect influence.

In contingency tables for the three species, the relief was opposed to wind velocity. These data are given in Table 20. A comparison of only the relief types frequented by the three species – the "all classes-row" in wind velocity is looked at in this first consideration – shows that ibexes were observed mainly in protected places, Klipspringer mostly in non-extreme (and exposed) and Geladas in exposed (and non-extreme) places. This distinction is highly significant (χ^2 : 200.17 df : 4; P $\ll$ 0.01). The comparisons of relief and wind velocity resulted in a significant pattern only for the Gelada baboon. When there was strong wind a preference for protected places is obvious. A tendency towards the same effect is visible in the table

Table 20. Contingency tables for the Walia ibex, the Klipspringer and the Gelada baboon in which local wind velocity is opposed to a summarized classification of the relief of the habitat within the 10 metres' range of the animals. The three classes of wind velocity are: 0=no wind, 1=high wind, which keeps grass and leaves moving, $\geq$2 strong wind (see Nievergelt 1966a). The table uses codified numbers of animals (see p. 68)

Species	Wind velocity	Relief of the habitat				Significance
		Exposed, edge ridge	Non-extreme	protected troughs	All classes	
Walia ibex	0	30	58	59	147	$\chi^2=9.07$
	1	110	354	449	913	df: 4
	$\geq$2	24	76	95	195	P$\approx$0.06
	All classes	164	488	603	1,255	
Klipspringer	0	18	19	15	52	$\chi^2=2.65$
	1	94	107	66	267	df: 4
	$\geq$2	15	28	16	59	n.s.
	All classes	127	154	97	378	
Gelada baboon	0	45	46	12	103	$\chi^2=39.04$
	1	177	147	116	440	df: 4
	$\geq$2	46	49	77	172	P$\ll$0.01
	All classes	268	242	205	715	

Table 21. Contingency tables for the Walia ibex, the Klipspringer and the Gelada baboon in which the local wind velocity (see Table 20) is compared with the exposure towards the wind. The summarized classification considers both the general exposure of the slope and the local topography within the 10 metres' range of the animals. The table uses codified numbers of animals (see p. 68)

Species	Wind velocity	Exposure towards the wind				Significance
		Exposed	Non-extreme	Protected	All classes	
Walia ibex	1	141	298	317	756	$\chi^2 = 4.32$
	≥ 2	20	69	78	167	df: 2
	All classes	161	367	395	923	n.s.
Klipspringer	1	65	84	83	232	$\chi^2 = 2.84$
	≥ 2	8	21	14	43	df: 2
	All classes	73	105	97	275	n.s.
Gelada baboon	1	144	196	65	405	$\chi^2 = 29.7$
	≥ 2	24	80	50	154	df: 2
	All classes	168	276	115	559	$P \ll 0.01$

for the Walia ibex, whereas the Klipspringer does not seem to respond to wind velocity in habitat selection.

A similar comparison is shown in Table 21. In this table wind velocity is compared with exposure towards the wind, a factor in which the wind direction and the local topography are considered. Of course the wind velocity class 0 does not occur in this comparison, as a situation without wind does not permit the exposure towards the wind to be stated. As in Table 20, only for the Gelada baboon was there a significant effect resulting. This effect coincides with the observation of Kawai (1979), that Geladas were huddled in the hail or sometimes took refuge in rocky holes or under overhanging rocks (Kawai and Iwamoto 1979). It also agrees with the finding that Geladas in the Zurich Zoo lower their activity and search for sheltered places in cold temperatures (V. Hauser, pers. comm. 7.5.1979). Many ungulates, and particularly goats, are well protected against cold (Hensel 1955).

Regardless of the degree of perfection of insulation against cold temperatures, because of the enormous differences in temperature that occur between sun and shade, it is to be expected that a large mammal will respond to this factor in its habitat selection (see Chap. 2, p. 11). A comparison with the relief, however, would be heavily biased, as topographically protected places are more likely to be in the shade anyway. In Table 22 a comparison is made between full sun, low sun, and shade on one side with the general exposure towards wind on the other. In contrast to Table 21, in order to exclude the mentioned bias in this wind exposure factor, local topography was not considered. As Table 22 shows, places exposed to wind were visited more often by all the species, if they were in the sun. The pattern is particularly obvious in the Walia ibex and the Gelada baboon. In a comparison between the species, the table also shows that ibexes were observed most often in shade, Klipspringer in partial sun, and the Geladas in full sun. This last result must be seen in connection with Table 20, where results show that Geladas frequent

Table 22. Contingency tables for the Walia ibex, the Klipspringer and the Gelada baboon in which the general exposure of the slope towards the wind is compared with sun or shade at the place of the animal. The table uses codified numbers of animals (see p. 68)

Species	Sun, shade	General exposure towards the wind				Significance
		Exposed	Non-extreme	Protected	All classes	
Walia ibex	In full sun	115	76	48	239	$\chi^2 = 81.81$
	In low sun	75	78	84	237	df: 4
	In shade	73	150	200	423	$P \ll 0.01$
	All classes	263	304	332	899	
Klipspringer	In full sun	23	28	10	61	$\chi^2 = 9.52$
	In low sun	30	63	42	135	df: 4
	In shade	15	36	26	77	$P = 0.05$
	All classes	68	127	78	273	
Gelada baboon	In full sun	111	141	32	284	$\chi^2 = 49.13$
	In low sun	55	56	46	157	df: 4
	In shade	41	30	47	118	$P \ll 0.01$
	All classes	207	227	125	559	

Table 23. Contingency table for the Walia ibex in which the relief of the habitats within the 10 metres' range of the animals is compared with the activity. The values are given in percentages. The 16-field test, which is based on the original numbers of observed ibexes resulted in: df: 9, χ^2: 85.24, $P \ll 0.01$

Activity of the ibexes	Relief				Total number of animals
	Exposed on ridges resp. edges	Non-extreme		Protected, mainly troughs	
		Ridge but on protected terrace	In slope		
Lying	18.8	18.2	17.6	45.5	346
Standing, not feeding	11.3	13.7	17.6	57.4	336
Feeding	9.9	7.1	30.5	52.5	876
Walking, not feeding	10.0	8.3	20.0	61.7	180
Total number	208	186	423	921	1,738

exposed places, a behaviour that would be difficult to understand as such when one recalls that this primate has to resist the often cold afroalpine climate. All data shown in Tables 20, 21 and 22 stress the direct influence of relief and weather conditions on the habitat selection via thermal comfort, particularly in the ibex and the Gelada. In Table 23 for the Walia ibex only, the relief is compared with the activity of the animals spotted. For an easier comparison, the values are given in percentages, but the significance is based on the original values in the 16-fields-contingency table. The following effects visible in Table 23 are of primary interest: Resting animals obviously have a preference for ridges; these are places which enable the animals to overlook the terrain. Feeding animals also show a behaviour which seems to be quite distinct from the other activities. The high numbers of animals on plain slopes is particularly evident. This pattern seems to be caused by the vegetation and is thus an indirect relief factor.

11.5 Relationship Between the Gradient, the Vegetation and the Group Size in the Walia Ibex, the Klipspringer and the Gelada Baboon

For any species the frequency distribution of groups of different sizes is most probably influenced by a number of different selective pressures (Wilson 1975; Bertram 1978). These pressures arise from ecological factors and social interactions such as the distribution and availability of food, the detection and avoidance of predators and the reproductive behaviour.

Groups of most Caprinae and many other ungulates are open, and therefore animals join the group and leave it again. For such species we can expect the group size to respond to factors which are of a limited character. Apart from the whole population size, which affects herd size (Schaller 1977) and apart from changes associated with the reproductive cycle (Nievergelt 1974), it has been observed in the Axis deer and the Zebra that the groups are smaller when the preferred food is sparse and scattered (Schaller 1967; Klingel 1967). The largest aggregations of ungulates can be observed in open grassland. Examples are Wildebeest, Eland, Oryx and Zebra in East Africa (Lamprey 1963; Talbot and Talbot 1963; Estes 1966; Schaller 1972; Leuthold 1977).

Species with closed groups such as the Klipspringer are more casual in maintaining their optimal group size. However, as in species with open aggregations, group size must be seen within the social and ecological context.

Within the three species compared it must be asked if group size varies in different types of habitat. The average grouping pattern has been presented in the introductory chapters to the three species (see Chap. 4.3, p. 35 for the Walia ibex, Chap. 5, for the Klipspringer, p. 39, and for the Gelada baboon, p. 42). The numbers of males and females in groups, as well as the numbers of single animals in the Walia ibex and the Alpine ibex were discussed in a comparative study on the two ibex subspecies (Nievergelt 1974). For the Walia ibex, the Klipspringer and the Gelada baboon, it can be said that group sizes remain at a similar level throughout the year. Some slight changes in the two ungulate species coincide with the reproductive cycle.

Table 24, which consists of various cross-tables, records for each of the three species the number of lone animals and groups which were observed in the different classes of the vegetation and the gradient, but separately for different sizes of groups (the same frame of cross-tables will be used for a comparison of the niches of seven mammals, see Table 25, p. 112).

As can be seen immediately, the range of existing group sizes is different in the three species. Group size varies most and reaches high numbers in the Gelada baboon; in contrast it is small and remarkably stable in the Klipspringer. This fact needs to be considered if comparisons are to be made between the species. However, comparisons within each species are the main interest. Chi-square tests were calculated for the summarized effects of the gradient and the vegetation separately. Thus the sums of the columns and of the rows were compared between group-size classes only. This restriction was expedient in order to avoid too low expectancy values in the test. The result is given at the bottom of the table.

Looking at the whole of Table 24, as well as at the level of significance, we can see: differences in habitat selection between group-size classes are very small or even non-existent for the two ungulate species but are fundamental in the Gelada baboon. In this species the table reveals the phenomenon that smaller groups occur mainly in steeper and more covered areas, whereas large herds aggregate particularly in plain and open areas stocked with Giant Lobelias. In most of the groups recorded, the animals were feeding, the major activity of the Geladas in the daytime (Iwamoto 1979). However, the daily activity pattern apparently promotes the result shown in the table. According to this pattern Geladas roost at night on the precipice, in the morning they climb up towards the plateau (Kawai 1979). This is the range where herds may join temporarily to multi-herds (see p. 42). Ohsawa (1979, p. 52) states this phenomenon even more generally: "It is unquestionable that small groups have a tendency to be attracted and drawn together into a large group on the plateau". Within the main study area between Gich Abyss and Amba Ras large herds of Geladas were only observed on the open plateau and feeding. This coincides also with the findings described and discussed first by Crook (1966), but more comprehensively by Dunbar and Dunbar (1974b, 1975), Dunbar (1977a), who carried out their observations partly in the same area, but to a greater extent at the western border of the Park in the Sankaber region.

As explained above (Chaps. 8.1 and 8.7, pp. 68, 75) the stepwise multiple regression which was carried out did not consider this obviously important relationship between gradient, vegetation and group size. This is one of the reasons why a theoretical distribution pattern was not calculated for this species as it was for the Klipspringer and especially the Walia ibex as the major object of this study.

Much more subtle differences in the pattern of habitat selection with increasing group size are indicated for the Walia ibex. In the group class of 8 to 31 animals, which includes the largest Walia associations observed, the following deviations from the pattern set by the two other classes can be seen: there is a less marked preference for steep slopes above 45°, but at the same time plain areas seem to be more definitely avoided. There is as well a greater concentration on areas with more dense cover with at least some shrubs or trees present. A Chi-square comparison of the data of group-size classes 1 (lone animals) and 2 to 7 only turned out to be not significant (for steepness: $\chi^2 = 5.2$; df. 3; n.s.; for vegetation: $\chi^2 = 6.5$; df. 3; P < 0.1 n.s.). In contrast, with the comparison of the two classes 2 to 7 and 8 to 31 individuals, a significance level was reached (for steepness: $\chi^2 = 7.4$; df. 2 – pooling due to low expectancy values –; P < 0.05; for vegetation: $\chi^2 = 9.6$; df. 3; P < 0.05). Thus the indicated habitat difference, when comparing the three group-size classes seems to be caused mainly by the habitat preference observed for the largest herds. This result does not coincide with the already mentioned and more commonly observed phenomenon that in open terrain the aggregation of larger groups is promoted. It is possible that this behaviour of the Walia ibex must be interpreted as a result of more frequent human disturbance in open areas. In the data shown in Table 24, a (not significant) tendency seems to be visible which suggests that lone

Table 24. Frequency in which groups of Walia ibex, Klipspringer and Gelada baboon were recorded in different habitat types, separately for various group sizes and lone animals, which are listed as group size 1. In the cross-tables the following classes were distinguished: For the vegetation: open, open with Giant Lobelias (Lob), single shrubs or trees (Sh./Tr.), open savanna and forest (For). For the gradient: <15°, 15°–30°, 30°–45° and >45°. With Chi-square, the frequency in the sum rows for the gradient and in the sum columns for the vegetation were tested. In the Gelada baboon, the first two and the last two tables were pooled for this test. Chi-square, degree of freedom and resulting probabilities of error are given below

Group size	Walia ibex Vegetation					Klipspringer Vegetation					Gelada baboon Vegetation				
	Open	Lob.	Sh./Tr.	For.	Sum	Open	Lob.	Sh./Tr.	For.	Sum	Open	Lob.	Sh./Tr.	For.	Sum
1 Gradient															
<15°	4	5	3	–	12	3	6	2	1	12	–	–	–	–	–
15°–30°	2	5	9	10	26	1	2	2	10	15	–	–	–	–	–
30°–45°	5	1	25	24	55	1	3	11	15	30	–	–	–	2	2
>45°	32	–	72	48	152	1	–	8	3	12	1	–	2	3	6
Sum	43	11	109	82	245	6	11	23	29	69	1	–	2	5	8
2– 7 Gradient															
<15°	4	3	2	1	10	7	12	4	6	29	1	–	1	–	2
15°–30°	2	13	4	9	28	3	9	8	18	38	–	1	–	4	5
30°–45°	8	8	29	50	95	4	13	23	28	68	2	1	6	11	20
>45°	62	6	104	87	259	7	–	17	15	39	11	–	25	13	49
Sum	76	30	139	147	392	21	34	52	67	174	14	2	32	28	76

Group size	Gradient	Walia ibex Vegetation					Klipspringer Vegetation					Gelada baboon Vegetation				
		Open	Lob.	Sh./Tr.	For.	Sum	Open	Lob.	Sh./Tr.	For.	Sum	Open	Lob.	Sh./Tr.	For.	Sum
8– 31	<15°	—	—	—	1	1						7	7	3	2	19
	15°–30°	—	2	1	—	3						1	4	2	3	10
	30°–45°	—	—	10	10	20						—	1	9	11	21
	>45°	2	—	14	7	23						6	1	11	7	25
	Sum	2	2	25	18	47						14	13	25	23	75
32–127	<15°											1	10	2	1	14
	15°–30°											1	1	1	—	3
	30°–45°											1	—	4	1	6
	>45°											—	—	1	2	3
	Sum											3	11	8	4	26
≧128	<15°											1	9	2	—	12
	15°–30°											—	—	—	—	—
	30°–45°											—	—	—	—	—
	>45°											—	—	—	—	—
	Sum											1	9	2	—	12
Significance		χ^2	df	P			χ^2	df	P			χ^2	df	P		
	Gradient	13.7	6	<0.05			0.8	3	n.s.			74.8	6	≪0.01		
	Vegetation	14.2	6	<0.05			1.2	3	n.s.			47.0	6	≪0.01		

animals were observed more frequently in level areas. Such a tendency is obviously caused by the males among the lone animals; these contribute 126 out of a total of 245 lone animals. Males, in comparison with females, do show a definite tolerance for less steep areas (see p. 147). Among the 12 lone animals seen on slopes of less than 15°, 11 were males.

For the Klipspringer no change of habitat is visible in comparing the two group-size classes. This fact is not surprising for two reasons: Firstly, the overall difference in group sizes is very small. The highest number recorded was a group of five animals standing together on one occasion. A comparison of the two analogous group-size classes in the Walia ibex did not result in a significant difference either. Secondly, it is quite possible that some of the single animals were in fact members of a group of two or three animals, but the other animals were simply not visible. For a small ungulate species which is perfectly able to hide, even in tall grass, it is easy to miss individuals. The number of single-animal groups (69 out of 243) in comparison with the data of Dunbar and Dunbar (1974a), who recorded 45 out of 129 for first sightings, but only 5 out of 40 in the prolonged sightings, indicates that such a bias has to be considered.

11.6 Observation Success at Muchila Afaf-, Kedadit- and Saha-Observation Points

As the multiple regression presented in the previous chapter is based on the observation data from Muchila Afaf, Kedadit and Saha, these observation points were considered in the analysis as a further environmental factor. In the final equation for the Walia ibex, the Saha observation point is included with a strong positive contribution (see Table 15). This result requires some explanation.

Obviously the factor observation point represents some otherwise overlooked environmental characteristics of the geographical unit as a whole, or certain factors which are particularly evident in a unit. It is possible that space as such, an unbroken and extended area which is suitable for ibexes, may be important. Such large-scale characteristics were ignored, because in the equation only the range of the hectare fields and their adjacent fields were considered. Evidence for the importance of the space stems from the theoretical distribution pattern, based on the regression equation and given and discussed further below (Chap. 14). In this pattern we can detect a number of hectare fields within the plateau range, which have received a high value according to their field characteristics but which are nevertheless unsuitable, because they are too isolated from a larger suitable area of habitat. A comparison of the geographical units of Muchila Afaf/Kedadit and Saha shows that this last place is situated in an altogether more extended area, with richer topographical relief at all levels of altitude. Apart from the map in Fig. 23 (p. 56), the photographs in Fig. 30, but also Figs. 2, 5 and 8 (pp. 8–16) show the characteristic appearance of this geographical unit. The foremost reason for the positive coefficient of Saha observation point therefore may be possibly that such an aspect of habitat which would be based on a larger scale may be of influence.

A second possibility may be that the area overlooked from Saha OP was actually overestimated in comparison with the geographical unit Muchila

Fig. 30. The geographical unit which refers to the observations made from Saha observation point. The rich topographical relief is evident

Afaf/Kedadit. According to the regular counts carried out from fixed observation points as from 1968 until 1977 there is a distinct decrease for Saha but only a slight decrease with a subsequent tendency towards recovery for Muchila Afaf/Kedadit (see Table 14, p. 80). This coincides with the impression of various observes that Kedadit, which was originally a point with an extremely variable observation success, turned out to be a fairly good place from which to show Walias to visitors. This tendency has to be seen in association with further facts. In the Muchila Afaf-Kedadit unit I have spotted comparatively far fewer males of the oldest age class than in the Saha unit and the relative number of kids observed was lower (see Table 7, p. 60). The number of old males increased after 1969; thus the near-absence of old males is not due to a permanent environmental situation (these changes in age distribution over the years will be presented in a different paper). This indication leads to the presumption that poaching activity – although generally low in the whole study area in 1968/69 – could have been relatively greater in Muchila Afaf/Kedadit. Old males with their heavy bodies and large horns are particularly desirable as prey (see also p. 27 and Fig. 13). It is perhaps not accidental that the only snare I found during the field period was on a rock band below Kedadit observation point. It seems therefore that the positive coefficient of Saha OP in the equation for the Walias is probably caused by two reasons: the rich quality and extension of the overall terrain in the Saha area and the possibly higher degree of disturbance in the Muchila Afaf/Kedadit unit in 1968/69.

11.7 Conclusions, Comparisons with the Alpine Ibex and Considerations on a Possible Tradition

With a regression analysis for the Walia ibex, the Klipspringer and the Gelada baboon, a general and average pattern of their habitat was presented in Chapter 10. In Sections 11.1 to 11.6 of this chapter, some supplementary and more detailed information on the ecological behaviour of these mammals was presented, together with some so far neglected environmental variables such as the observation factors (see Chap. 8), the different seasons and times of the day, and the weather conditions.

As mentioned already in Chapter 8.7 it is very likely that most of these differentiations have led to an increase in the variation in the data, and consequently have probably lowered the degree of significance of the regression equation, but have not biased the result heavily. With regard to seasonal and daily effects, the most apparent are the distinctive preferences for certain compass directions as shown in Figs. 26, 27 and 28. The three main observation points G3, G4 and E2 on which the regression analysis is based have been visited each month, and the seasons are fairly equally represented (Table 5, p. 58). Thus the equation is a good estimate for the yearly average and also for those factors that may cause a seasonal effect. In contrast, the times of day were given unequal attention; the nights had to be ignored and within the day, the morning hours were used as the main data-collecting period (see Chap. 7.2). The regression equation is therefore to a large extent a yearly average of morning situations. As indicated by Fig. 26, if unbiased, observations with respect to a balanced morning–evening pattern were achieved, the generally east-facing slopes would lose some weight (but still remain positively correlated) and simultaneously the west-directions would be favoured (but remain negatively correlated).

So far the habitat selected by the Walia ibex has been compared with the overall environmental situation in the study area and with the habitats of the two potential competitors living in the same area, the Klipspringer and the Gelada. This synoptic view was applied to work out the particular niche of the Walia ibex. In order to discuss the question of which aspects of the ecological behaviour of this member of the Caprinae have been retained and which aspects the afroalpine environment has changed, a summarized comparison with the Alpine ibex is given subsequently. For this consideration of the importance of various environmental factors for the two ibexes, I shall refer to Nievergelt (1966a, 1974), Hofmann and Nievergelt (1972) and Schaerer (1977).

Alpine and Walia ibexes show definite preferences for steep slopes. Two further factors, the altitude and the compass direction of the slope, are obviously also important for both species or subspecies. In entirely different climatic situations, however, the response cannot be the same as it is for the gradient of the slope, but in both ibexes there are clear indications of preference for thermal comfort (for example in Fig. 26) and for the vegetation (Fig. 27). The same relation was found in the main in the comparison between sunny and shady places and topographical exposition (Table 22, Nievergelt 1966a, p. 69). Further similarities and differences are listed with reference to the different Walia classes in Chapter 16.7. Among those differences, not obviously caused by the very different climatic situation (see Chap. 2), the most remarkable ecological peculiarity for the Walia ibex compared with the Alpine ibex is the strong utilization of shrubby, savanna-type and forested habitats (see also Nievergelt 1974, p. 327, Table 1). In the Alpine ibex a similar behaviour is known only from colonies in the Prealps where the alpine zone is poorly represented or lacking (in Switzerland for instance in the Augstmatthorn and the Justistal). The behaviour of the ibex in these areas was interpreted as using the closer vegetation cover to escape the heat of the day. However, these are clearly exceptional cases, and the amount of forest is definitely greater for the Walia ibex even in comparison with these unusual colonies of the Alpine ibex. One possible key to the understanding of this behaviour of the Walia ibex may be seen in the postulated immigration period when the ibex moved into the Simen. At that time, the vegetation belts were lower (see Chap. 4.2, p. 34); the zone of the rocky escarpment must have been well above the timberline. With the changing climate, and the approaching forest, the Simen ibex had almost no possibility of moving upwards with its habitat because of the characteristic shape of the mountains with soft plateau and hills at the top, and because of the limited altitude of the rocky parts in wide areas (see Fig. 3, p. 9 and Table 11, p. 71). An adaptation to partial forest life seems therefore readily understandable. An additional or different reason may be the light character, at least of the upper forest in the Ericaceous belt with its small trees. The search for shade temperatures and soft vegetation, mainly during the dry season, may have steered the Walia ibex towards forested habitats.

The preference of the Walia ibex for comparatively dense vegetation remains also a remarkable peculiarity, if the general character of the Caprinae as glacier-followers is considered (see p. 33, as well as Geist 1971a). Having in mind the pioneerlike nature of the Caprinae, we could well expect that the environmental changes described in Simen, mainly the opening up of the vegetation by man, have actually improved the habitat for the ibex. However, the data on the distribution presented later (Chap. 14) and on its habitat shown in this investigation do not support such a theory. The apparent preference of the animal for fresh food is a further reason for its rejection (Chaps. 11.1 and 16.6). Nevertheless, the following rather speculative hypothesis remains. It is possible that the ibexes in Simen select and maintain their largely wooded habitat to a greater or lesser extent due to tradition. Such a possible partial tradition may have originated when the Simen became more and more forested due to climatic changes, but it remained when man began to clear forested areas, but kept these open places under control. It is plausible that pursuit of the animals in open terrain undoubtedly favours and may help to preserve such tradition. It must be stressed however, that the open zone during the time of the presumed invasion of the ibex into the Simen mountains was in fact a zone above the timberline and thus a definitely colder habitat than the present man-made open areas. It may be presumed that the absence of heat-protection would have to be compensated for by the relief, e.g., gorges and other shady places if the covering trees are absent. But of course it is unknown, and according to the observed and here presented habitat selection, even rather unlikely, that the Walia ibex would in fact tolerate and occupy some of the impoverished and so far neglected ranges.

In order to judge the idea that tradition might play a role of decisive importance in maintaining habitat selection it is important to consider evidence in other, closely related species and in the social system of the Walia ibex (discussed below, Chaps. 15 and 16). Long-lasting traditions have been postulated by Geist (1971b) for the Bighorn sheep that is migrating on routes through forested valleys that were apparently established between deglaciation and forestation. According to Tuercke and Schmincke (1965), introduced, but genetically pure Mouflon populations on the European continent inhabit exclusively forested areas. In such populations however, when sheep individuals were added that did not select forests as the only habitat, the whole population began to occupy open areas as well and thus changed their original habits successively. Personally, I do not see a more plausible interpretation than tradition to explain this phenomenon. This example would be similar to the Walia ibex in the Simen in so far as due to tradition a nearby habitat was at first neglected. In the Alpine ibex we may find some evidence for learning and presumably tradition, for instance in the history of the population in the Safiental, e.g., the regular migrations to the winter ranges and in the population of the Swiss National Park, where it lasted almost 20 years until the ibexes had occupied the ideal winter range in Chanels as from their adopted place in the nearby Val Cluozza, where the first introductions took place (Nievergelt 1966a).

A further comparison with the Alpine ibex might give some support to the tradition theory in the Walia ibex. Unlike the Chamois (Krämer 1969) and the Red deer (Schloeth and Burckhardt 1961) the Alpine ibex shows an apparent inertia in occupying new, adjoining ranges. However a small tendency to a further dispersal does exist. There is much evidence that the "exploring class" are the males and that the "exploring period" corresponds to the non-rutting seasons from spring to autumn. Dispersing males were observed, e.g., in the following populations: Safiental, Augstmatthorn, Justistal, Flims (Nievergelt 1966a). In the Walia ibex, where a period with no rutting activity is lacking and where the males' tactic appears to include regular visits in the female groups (Chaps. 15 and 16), there is no similar period where the males were able to explore further ranges without decreasing their chances of meeting a female in oestrus. Tradition may therefore be important for the Walia in the frequenting of its habitats. However, the hypothesis that tradition would actually paralyse the occupation of an ecologically different, but possible and geographically adjoining habitat appears unlikely or at least doubtful. As this open question obviously reveals a number of practical aspects also, it might be considered a tempting and useful experiment to try carefully to build up an adaptation to treeless habitat – the protection being provided by rocks only – and to rough food in enclosed ibexes. The experiment could be combined with the project to establish a breeding nucleus of the Walia ibex in captivity. The release of such trained and individually marked ibexes should be planned accordingly in rocky, but open areas, e.g., near Bwahit mountain. Of course, prior to such a release regular observations and an optimal control would have to be organized. It is expected that through this experiment an answer to the question, whether this African ibex – in adapting towards montane forests, tree heather vegetation and a tall, fresh mountain steppe – has genetically lost or modified the general "Capra order" to follow glaciers and forage in rough vegetation or if this order was simply forgotten.

12 Separation in the Habitat of the Walia Ibex, the Klipspringer, the Gelada Baboon, the Bushbuck, the Colobus Monkey, the Simen Fox and the Golden Jackal

The main characteristics of the habitat of the Walia ibex and that of the two further large mammals of the area, the Klipspringer and the Gelada baboon, as possible competitors of the ibex, were described in Chapters 10 and 11. In this chapter considerable emphasis is also given to these three, mainly spotted species, but, in order to widen the field of comparison, four additional mammal species are also examined.

Two of the main factors which are appropriate to differentiate the specific niche of the species within the given environment are the gradient of the slope and the vegetation (see the Tables 15–17 as well as 19, 24 and Fig. 29). In Table 25, in cross-tables for each species, the codified number of animals is given for each class combination of the two environmental factors, gradient and numbers of shrubs or trees. In this table, each of the seven mammals listed in the title is considered; behind the name of the species – as a further differentiating factor – the arithmetic mean of the altitude is given (this value is biased as explained above, p. 97). In Table 25d for easy comparison the rank order or relative frequencies among the three main species Walia ibex (W), Klipspringer (K) and Gelada baboon (G) is indicated by the sequence of the letters written. The table shows the following pattern. The three main species listed in the first two lines are differentiated only around 170 m in the average altitude. The given value is within the actual escarpment range. The relatively well-frequented habitat classes are: for the Walia ibex, steep slopes, open vegetation and single trees and shrubs only, for the Klipspringer moderate gradient, open vegetation with Lobelias and open to dense forest, for the Gelada baboon steep as well as flat terrain and open vegetation. These data coincide generally with the percentages given by Dunbar (1978a) as frequenting escarpment, ridge top (plateau) and gorge. If the biomass or the directly counted numbers of animals were presented instead of the codified numbers in Table 25 as in Dunbar's study, the value for the number of Geladas remaining in flat areas would total around half of all the animals, as the largest herds were seen on the plateau. This effect was shown in Table 24, which consists of analogous cross-tables as presented here in Table 25, but differentiated for various group-size classes.

The two carnivores, the Simen fox and the Golden jackal, mainly frequent the high plateau, as shown by the average-altitude value and by the clear preference for flat or level areas, particularly those stocked with Giant Lobelias. This is actually the characteristic habitat of an important prey, small mammals, the main prey at

Table 25. Ecological separation of seven mammal species in Simen according to the gradient of the slope and the type of vegetation. Four classes of gradient were distinguished and the following vegetation classes: open vegetation (open), open with the Giant Lobelias *Lobelia rhynchopetalum* (Lob.), single shrubs and/or trees (Sh./Tr.), and open, savanna type forest and forest (For.). In seven tables, the codified numbers of observed animals are given for each of the classes. Additionally the arithmetic mean of the altitude is listed for each of the species. In table d the rank of the relative frequencies between the Walia ibex (W), the Klipspringer (K), and the Gelada baboon (G) is indicated by the sequence of the letters

(a) Walia ibex; 3,354 m

Gradient	Open	Lob.	Sh./Tr.	For.	Sum
< 15°	12	12	8	6	38
15°–30°	7	43	24	29	103
30°–45°	23	20	131	183	357
> 45°	191	13	384	280	868
Sum	233	88	547	498	1,366

(b) Klipspringer; 3,518 m

Open	Lob.	Sh./Tr.	For.	Sum
19	30	10	13	72
8	20	18	46	92
9	31	57	73	170
16	0	43	35	94
52	81	128	167	428

(c) Gelada baboon; 3,488 m

Gradient	Open	Lob.	Sh./Tr.	For.	Sum
< 15°	49	170	44	15	278
15°–30°	11	29	14	24	78
30°–45°	13	8	82	87	190
> 45°	61	5	128	79	273
Sum	134	212	268	205	819

(d) Rank in frequency of W, K, and G

Open	Lob.	Sh./Tr.	For.	Sum
G K W	G K W	G K W	K G W	G K W
K G W	K G W	K W G	K G W	K G W
K W G	K W G	K G W	K W G	K W G
W G K	W G K	W G K	W G K	W G K
W G K	G K W	W G K	K W G	–

(e) Simen fox; 3,605 m

Gradient	Open	Lob.	Sh./Tr.	For.	Sum
< 15°	0	13	2	1	16
15°–30°	0	3	1	0	4
30°–45°	0	0	0	1	1
> 45°	0	0	0	0	0
Sum	0	16	3	2	21

(f) Golden jackal; 3,675 m

Open	Lob.	Sh./Tr.	For.	Sum
2	12	5	0	19
0	4	2	1	7
0	0	1	0	1
0	0	0	0	0
2	16	8	1	27

(g) Colobus monkey; 2,750 m

Gradient	Open	Lob.	Sh./Tr.	For.	Sum
< 15°	0	0	0	0	0
15°–30°	0	0	0	8	8
30°–45°	0	0	1	10	11
> 45°	0	0	1	2	3
Sum	0	0	2	20	22

(h) Bushbuck; 2,790 m

Open	Lob.	Sh./Tr.	For.	Sum
0	0	0	6	6
0	0	0	0	0
0	0	3	0	3
0	0	0	0	0
0	0	3	6	9

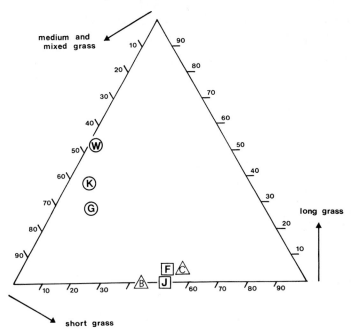

Fig. 31. Habitat of seven mammal species according to the type of ground vegetation: Walia ibex (*W*), Klipspringer (*K*), Gelada baboon (*G*), Simen fox (*F*), Golden jackal (*J*), Bushbuck (*B*) and Colobus monkey (*C*). In the *triangle coordinates* one can see the amount in percentages of the (codified) number of animals spotted in long-grass vegetation (particularly *Festuca macrophylla, Poa simensis, Carex monostachya, Koeleria convoluta*) in medium and mixed-grass vegetation (mainly *Festuca abyssinica, Danthonia subulata, Poa simensis, Pentaschistis pictiglumis*) and in short-grass vegetation (mainly *Poa simensis, Danthonia subulata, Festuca abyssinica, Agrostis* sp., *Swertia* sp., *Merendera abyssinica*). The species mentioned are valid for higher altitudes only, approximately above 3,300 m; thus not for the situation of the Bushbuck and the Colobus. The *different symbols* were used to underline which animals are ranging in ecologically similar habitats. In the figure, the Grass rat *Arvicanthis abyssinicus* would have to be drawn presumably very close to the *F* and *J*

least for the Simen fox. In this area the first small mammal species worth mentioning is the daily active Grass rat *Arvicanthis abyssinicus*, which lives in large numbers in some areas (see also p. 44 and Müller 1977). The second primate listed, the Colobus monkey or Gureza lives in Simen like the Bushbuck in the montane forests of the lowlands.

In Fig. 31 the type of ground vegetation for the seven species is shown. In triangle coordinates, the numbers in percentages of animals seen in long-grass vegetation, in medium- and mixed-grass vegetation, and in short-grass vegetation is shown. Among the three main species, as shown also in Fig. 29, the *Festuca-macrophylla* long-grass vegetation is used mainly by the Walia ibex, whereas the *Poa simensis/Danthonia subulata* short-grass vegetation is used mostly by the Gelada baboon. The short-grass vegetation is frequented most extensively by the two carnivores. The Bushbuck and the Colobus were drawn in the same region of the graph as the fox and the jackal, although their habitat in the montane forest is

Table 26. The number of animals of seven mammal species seen in places with up to one third and above one third ground vegetation coverage. The class with little vegetation is separated according to the characteristics of the open ground. A comparison of the frequency of the Walia ibex, the Klipspringer and the Gelada baboon in a contingency table resulted as: χ^2: 57.34; df: 4, P<0.01; thus the pattern is significantly different

Species	Vegetation coverage up to $^1/_3$; rest:		Vegetation coverage above $^1/_3$
	Rock	Sand/soil or mixed	
Walia ibex	145	117	1,112
Klipspringer	30	4	394
Gelada baboon	43	43	733
Simen fox	1	0	20
Golden jackal	1	0	26
Colobus monkey	0	0	27
Bushbuck	0	3	6

different. This occurred because the distinction was made only according to the length of the grass, and not by the vegetation units. Naturally because of the much lower altitude in the zone of the Olea-Maesa-Juniperus forest, the ground vegetation consists of different species.

As an observation factor, the amount of ground vegetation cover was registered for each group of animals spotted and recorded as zero, one, two or three thirds. For those parts of the place without vegetation cover it was noted if the ground was rock, sand or soil. In Table 26 for the seven mammal species, the codified number of animals seen in places with poor vegetation, where the coverage was estimated to average only one third or less, was contrasted with those animals seen in ground with more dense vegetation. The class with poor vegetation was subdivided, and a distinction made between rock and sand or soil covering the major part of the rest of the area. Comparing the Walia ibex, the Klipspringer and the Gelada baboon, the relative numbers of animals observed in the three classes were significantly different. Walias were in habitats with little vegetation cover more often than the other two species; the Klipspringer preferred a site well covered with vegetation, and it was particularly evident that this species did not seem to favour a terrain with open sand or soil. For the carnivores and the Gureza, there appears to be a preference for closer ground vegetation with a cover value of more than one third, although the figures are small. The results for the Walias and Klipspringers seem to contradict the findings given in Fig. 31. But this is false because the larger quantity of tall grass in the results for the Walia ibex may well be combined with areas that have low vegetation coverage.

In a final comparison of the habitats, the amount of open vegetation, shrubs and trees is opposed to the topographical relief. In Table 27 for each class combination, the rank order in the relative observed frequencies is given for the three species

Table 27. Comparison of the habitats of the Walia ibex (W), the Klipspringer (K), and the Gelada baboon (G) according to topographical relief and type of vegetation. The relief was distinguished into the classes: exposed, non-extreme, and protected; for the vegetation the same classes as in Table 25 were used. With the sequence of the letters W K G, the rank order in relative frequency of the three species is given

	Exposed			Non-extreme			Protected			Sum		
Open	**G**	K	W	**G**	W	K	**W**	G	K	G	W	K
Lob	**G**	K	W	**K**	G	W	**K**	G	W	G	K	W
Sh./tr.	**K**	G	W	**K**	W	G	**W**	G	K	W	G	K
Forest	**K**	W	G	**K**	W	G	**W**	K	G	K	W	G
Sum	G	K	W	K	W	G	W	G	K	—	—	—

considered. It is clearly seen that Walias mostly frequent protected terrain, Klipspringers exposed areas with denser vegetation and protected areas with open vegetation, and the Geladas were seen more often in open and exposed terrain. Similar data with reference to the relief are shown in Tables 20 and 21.

The data given in Chapters 10 to 12 all show that the habitats frequented by the three herbivores, Walia ibex, Klipspringer and Gelada baboon show clear differences, as well as similarities. The niches are different, but overlapping. The question as to what extent, if at all, the niche of each species is influenced by the ecological neighbouring species cannot be answered at present, as no areas were investigated with comparable methods, in which for instance only one of these species occurs. In an ecological comparison of 14 ungulate species in the Kyle National Park in Simbabwe/Rhodesia, but with no Caprinae among them, the Klipspringer showed an outstanding preference for rocky terrain (Ferrar and Walker 1974). A comparison of the very moderate preference for steep slopes shown by the Klipspringer in Simen with the above quoted result could indicate that the Klipspringer in Simen has displaced or reduced its niche towards smoother terrain due to the presence of the Walia ibex. However, this conclusion is clearly premature. The respective area in Simbabwe/Rhodesia is far less mountainous and the environmental factors were classified and computed in an approach that does not permit direct comparisons.

The overlap of the niches occupied is also confirmed by the number of hectare fields frequented by more than one of the three species. Out of the 664 Walia observations carried out within the main study area, in 235 cases the Walias were observed on a hectare field, on which Klipspringers (in 139 cases) or Geladas (in 150 cases) or both (in 54 cases) could also be recorded at least once in the study period. The 238 Klipspringer observations in the same area occurred on hectare fields that were also used by Walias in 99 cases, by Geladas in 54 and by both species in 37. The 180 Gelada observations were, in 91 cases, on fields also used by Walias, in 48 by Klipspringers and in 28 by both species.

In order to attain a deeper understanding of the relationship between these three herbivorous species we now ask two questions: How often were animals of these species observed, while they were together, using the same place, as compared with

Table 28. Associations between the Walia ibex (W), the Klipspringer (K) and the Gelada baboon (G). The number of observed associations (obs.) is compared with the expected value if the species are aggregating by chance only (exp.). The values are given for three observation points separately and for the whole main study area. For this whole range the coefficient of association is also added (Dice 1945)

Considered range	Associations between							
	W–K		W–G		K–G		W–K–G	
	obs.	exp.	obs.	exp.	obs.	exp.	obs.	exp.
OP Muchila Afaf	5	0.77	10	1.98	6	0.68	1	0.02
OP Kedadit	2	0.72	7	0.86	0	0.17	0	0.01
OP Saha	4	1.79	10	1.59	1	0.67	1	0.05
Whole main study area	17	6.64	37	6.59	8	2.79	2	0.09
Coefficient of association	2.6		5.6		2.9		2/0.09	

the probability of a random aggregation; and what was their behaviour, when members of different species met?

The number of expected associations between species, if these associations occurred at random, were calculated separately for each observation point and each hectare field and the values were added later. In the equation, w indicates the number of samples in which Walia ibexes occurred, k the number with Klipspringers and g the number with Gelada baboons; and n indicates the total number of samples for each observation point and thus the number of visits at this point. For each hectare field, the chance that, for instance, the Walia ibex and the Klipspringer occur together once in all the samples is therefore (see also Dice 1945):

$$w/n \cdot k/n \cdot n = \frac{w \cdot k}{n}$$

The chance that all three species occur together is:

$$w/n \cdot k/n \cdot g/n \cdot n = \frac{w \cdot k \cdot g}{n^2}$$

The summed values for all the hectare fields and the various observation points were afterwards compared with the number of aggregations actually observed. The result is shown in Table 28 for all the observation points of the main study area and also separately for the three major observation points. For the whole area the coefficient of association was also given (Dice 1945; Mc. Millan 1953). Table 28 indicates quite clearly that the species being considered were associated more frequently than would be expected by chance alone. Of course we must keep in mind that the units of the grid-pattern-system are as wide as one hectare and that the size of the fields applied affects the result. However, the result is good evidence of a peaceable disposition between those species. Apart from an imagined associating mechanism between these species, which will be discussed afterwards, the degree of

Fig. 32. Two Walia males of about 5 years passing a Gelada one-male unit; 29.8.1968 (see text)

association is also favoured by the following purely ecological reasons. (1) In Fig. 28 (p. 96) it was shown that for the compass direction of the slope the terrain seasonally favoured by the three species changes in quite a well-synchronized fashion. Such effects – it is very unlikely that the ecological synchronization would be a consequence of a possible interspecific social mechanism – of course increase the chance of occasional aggregations. (2) Disturbance of the animals in certain areas promotes aggregations in quiet zones. (3) Narrow passes which connect different areas favour occasional meetings of different species. I suspect, but cannot decide, that these ecological reasons may in fact explain the result of Table 28 to a satisfactory degree. In the discussion we must also consider the direct reaction of the animals of one species when meeting other species.

No agonistic behaviour was seen when Walia ibex and Klipspringer met. They were often seen peacefully feeding close to one another. On three occasions, the Klipspringers ran away, when Walias (males and/or females) moved – I presume accidentally – approximately towards them.

Between Walias and Geladas the situation is similar, but in every case when Walias advanced towards a group of Geladas, these impressive primates moved aside slowly but only a short distance. Twice I observed that a Walia actively chased away Geladas. In both cases the ibex was a yearling. This indicates that the character of the actions was presumably playful. In a further case, when two Walia males passed a Gelada group, a Gelada male seemed to respond with a copulation, which gave the impression of a hectic grasping for possession (Fig. 32). This reaction of the Gelada male seems to be analogous to the one described by Mori (1979, p. 185): "Leader males tend to copulate or groom following a quarrel with the leader of another unit, presumably as a demonstration and reaffirmation of their ownership and absolute monopolization of the females of their own unit".

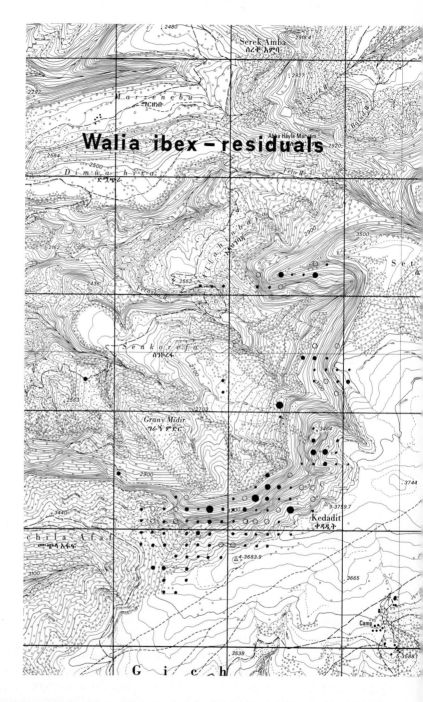

Fig. 33. Plot of the residuals of the regression equation referring to the Walia ibex and drawn on the topographical map of Simen. As in the multiple regression only hectare fields with a visibility greater than 20% were considered. As can be seen in the legend, in the fields with *small dots* the observed number of animals is of the same order as predicted by the equation,

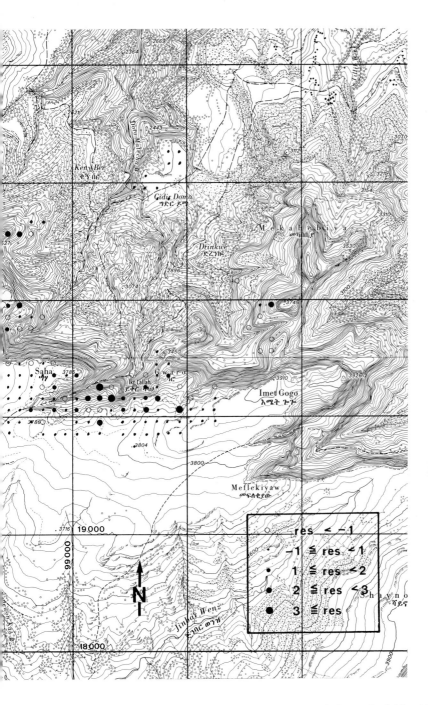

in fields with *open circles* the number of observed animals is lower, in fields with *large* and
solid dots the number is higher. The coordinates correspond to those used in the map of
Stähli and Zurbuchen 1978

No interaction was seen between Klipspringers and Geladas, even while foraging close to one another. In one case only, six Geladas moved to the side when two Klipspringers approached and then returned again to their original place, when the Klipspringers had passed. But in this example it is also possible that the movement of the Geladas was entirely independent of the Klipspringers.

This peaceful interspecific behaviour and particularly the giving way of the Geladas – at least while meeting the Walia ibex – is astonishing, because an adult Gelada must be physically superior to the ibex and the Klipspringer.

It is possible that the following considerations might be a key to understanding these observations and the coefficient of association given earlier. All three species are entirely herbivorous, but their preferred food differs considerably. This was shown by Dunbar (1978a) and is confirmed by my own data, as will be seen below. Assuming, therefore, little foraging competition, it might be advantageous for members of a species to have other animals nearby, as this situation supports their effort to avoid predators. A group of quietly grazing animals of another species can thus be used as an indicator of a predator-free zone. This mutual benefit may be particularly useful in a mountainous area, where a predator is able to approach its prey concealed by the relief and where the local wind direction often deviates unpredictably. Groups of different species may therefore use each other as a warning service, delivering information from an adjacent place and employing different sense organs. Clearly, all these considerations are only speculations, but they are also supported by the personal experience that it was almost impossible to approach a Walia ibex group in the presence of Geladas. The Walias clearly responded to the acoustic reactions of the Geladas. They became unsettled and attentive. Sometimes, they disappeared.

However, although these considerations about anti-predation may explain peaceful associations, they do not explain why the Geladas play an inferior role when meeting Walias. Is it possible to postulate that a dead animal attracts various predators and that this may be important in the context of this cautious Gelada behaviour? or, is it simply the more economic way and the shorter interruption for the Gelada to quickly give way and immediately concentrate again on feeding and on watching the activity of the many other members of the baboon society instead of threatening a harmless ungulate? As yet I do not have a satisfactory answer.

13 The Plot of the Residuals of the Regression Equation – a Test for Detecting Overlooked, but Decisive Independent Variables

In Figs. 33 (p. 118/119) and 34 (p. 122/123), drawn on the topographical map of the Simen mountains, the plot of the residuals is shown which result from the regression equations and which refer to the Walia ibex and the Klipspringer. For the Gelada baboon such a plot is not given nor a theoretical distribution pattern as presented in the next chapter, because mainly due to the major changes in habitat selection shown in Table 24 (p. 104), an average and overall pattern could not be satisfactory.

In Chapter 8.7, p. 76 it was explained that a clumped pattern of either positive or negative residuals would indicate the existence of an environmental factor so far ignored. In contrast, a random distribution would make such a missing factor unlikely. As Fig. 33 shows, the up and down deviations (residuals) seem to be scattered all over the area. One exception concerns an accumulation of too frequently visited fields on the north-facing slope opposite, and southeast of Saha observation point. The high values for those fields obviously stem from a nicely formed rock crevasse which was very regularly used by ibexes as a mid-day sleeping hide. The crevasse is fully visible from Saha observation point. The quality of those hectare fields near Saha is therefore based largely on a local factor which could not be included with the ordinary field factors. A small accumulation of negative residuals is visible in the area of Muchila Afaf observation point. The following explanation seems likely. I passed through or nearby these fields when I visited the three lookouts of Muchila Afaf observation point. Visibility of the fields and thus expectancy was therefore rather high. However, due to the fine tree heather forests in many parts of this area the sight-distance was relatively low. It is therefore possible that a remarkable number of animals disappeared from these fields without being spotted at the approach of the observer. Apart from these two places, the one near Saha, the other at Muchila Afaf, the pattern in Fig. 33 seems to be of random character. I therefore conclude that there is no need to look for the existence of any as yet unidentified major influencing factor, assuming that such a factor were important on the hectare-scale level of the field factors already applied. This limitation must be noted as the effect of space on a larger scale, as discussed in Chapters 11.6 and 14, cannot be seen in this figure.

In Fig. 34, which refers to the Klipspringer, the whole pattern does not seem to be clumped, and therefore the same conclusion as for the Walia ibex seems justified.

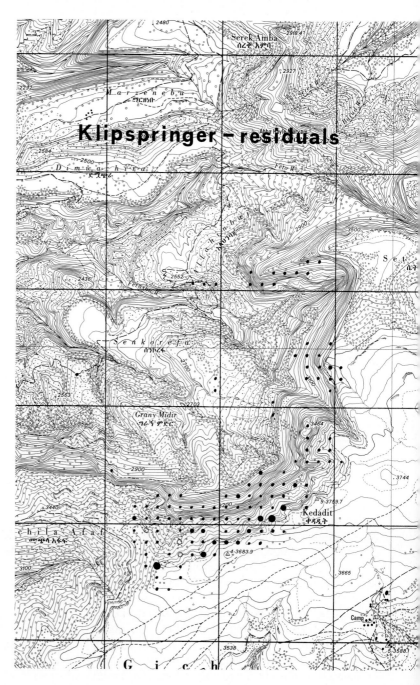

Fig. 34. Plot of the residuals of the regression equation referring to the Klipspringer. See for further information Fig. 33

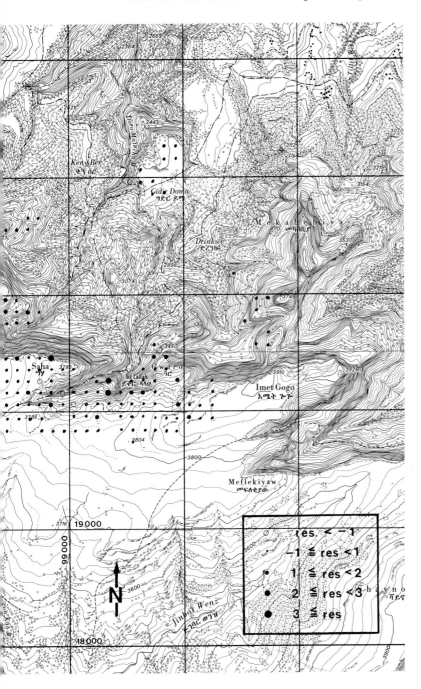

14 The Theoretical Distribution Pattern of the Walia Ibex and the Klipspringer Within the Study Area

The places directly recorded where animals of the two species were observed do not reveal the true pattern of distribution. Such a map with all the records collected in Simen is heavily biased because of two reasons: Firstly, I visited several zones regularly, (this concerns the three main observation points G3, G4, E2), some only occasionally, but certain zones remained that happened to be ignored. Secondly, as described earlier, within a geographical unit the visibility was unlike from hectare field to hectare field, for instance because of different vegetation cover, relief and/or observation distance. I therefore decided against using a map to show all the observations.

However, on the basis of the 5815 hectare fields being classified in the study area with the help of the field factors, and from the regression equation, which was generally confirmed by further results, it was possible to calculate a habitat-value for each of the hectare fields, and thus to plot a theoretical distribution pattern into the area. Thanks to the using of dummy variables in the multipe regression (see pp. 72 and 85), calculations were carried out on the level of the classes and not the original field factors. Therfore the classified hectare fields could be directly valued by the equation. The differing visibility of the hectare fields from the observation points were considered when the dependent variable for the described.and basing regression equation was determined. In contrast, for the calculation of the overall theoretical distribution the same visibility had to be applied for each field. This theoretical pattern or prognosis is not biased, but of course deviations from the true distribution do exist, even if due only to the fact that not all the variations in the data are explained by the regression equation. The two patterns for the Walia ibex and the Klipspringer, that may in fact be considered a habitat map for the two ungulate species, are shown in Figs. 35 and 36. For both species, the distribution presented coincides largely with the impression of the observer about the specific values of the areas included for the two species. For the Walia ibex the positive value of the escarpment range is obvious, particularly at those parts with many peaks and gorges. The generally negative values of the high plateau fields are also very clear. For the Klipspringer, the pattern is definitely more smooth, as might be expected because of its less pronounced ecological behaviour as described above.

As a test of the reliability of the model presented for all the valued hectare fields in the study area, which had not been considered in the regression analysis, a comparison was made between the predicted theoretical values of the hectare fields (the prognosis) and the number of animals actually observed. For both species,

Table 29. Comparison of the hectare fields which were not included in the regression analysis between the predicted value of each of those fields ($\hat{y}$) and the number of Walia ibexes and Klipspringers observed on the fields. The figures given in the contingency tables for the two species represent the numbers of hectare fields. Significance was examined with Chi-square test

Walia ibex (χ^2:141.0, df=8, p$\ll$0.01)

Codified number of Walias observed	Prognosis Predicted value for hectare field					
	$\hat{y} \leq 0$	$0 < \hat{y} \leq 0.75$	$0.75 < \hat{y} \leq 1.5$	$1.5 < \hat{y} \leq 2$	$\hat{y} > 2$	total
0	1,783	1,135	846	377	259	4,400
1 or 2	10	32	41	18	15	116
≥ 3	3	21	36	22	15	97
Total	1,796	1,188	923	417	289	4,613

Klipspringer (χ^2:41.3, df:6, P$\ll$0.01)

Codified number of Klipspringers observed	Prognosis Predicted value for hectare field				
	$\hat{y} \leq 0$	$0 < \hat{y} \leq 0.5$	$0.5 < \hat{y} \leq 1$	$\hat{y} > 1$	total
0	1,001	1,969	1,087	458	4,515
1 or 2	5	22	33	13	73
≥ 3	1	7	11	6	25
Total	1,007	1,998	1,131	477	4,613

separately, a contingency table is given in Table 29. It reveals with obvious significance that those hectare fields which are theoretically qualified as Walia- and as Klipspringer-habitats respectively, in fact were relatively more frequented by the corresponding species. This result is remarkable, because it is based only on those fields which had been excluded in the multiple regression. For the Walia ibex only one further test could be carried out in comparing the Walia observations made by Dr. H. Hurni during his Park-Warden-ship in 1975 and 1976 with the prognosis shown in Fig. 35. This comparison is given in a similar contingency table (Table 30). The significant result of this test also indicates that the calculated prognosis can be considered reasonable.

Fig. 35. Theoretical distribution pattern of the Walia ibex within the study area according to the regression equation. As symbols, in the sequence of increasing values of the hectare fields, *open circles* were applied and *small dots* up to *large dots* (see legend). The same five classes were used in Tables 29, 30 and 31. The distribution pattern may be considered a habitat map of the Walia ibex. The coordinates correspond to those used in the map of Stähli and Zurbuchen 1978

Walia ibex – habitat – map

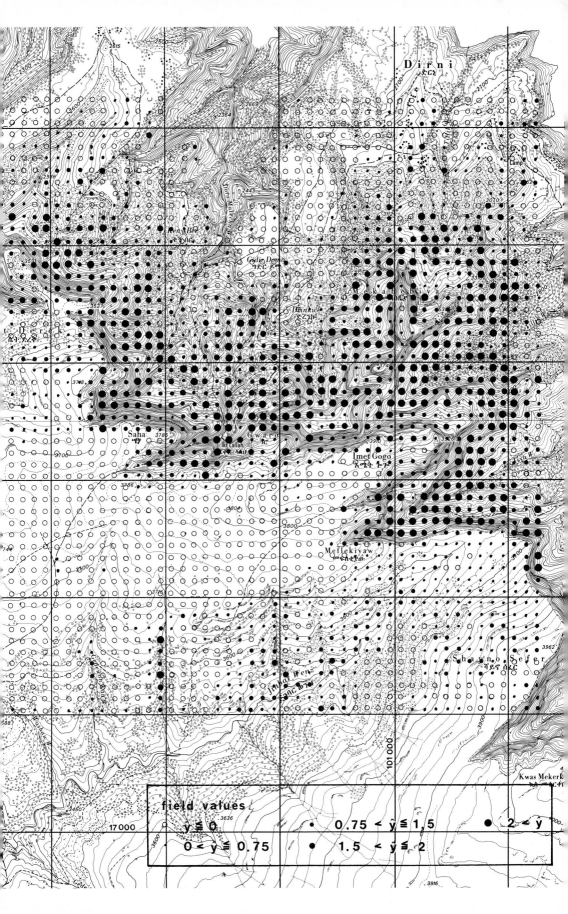

field values

○	$y \leqq 0$	•	$0.75 < \acute{y} \leqq 1.5$	●	$2 < y$
○	$0 < y \leqq 0.75$	•	$1.5 < \acute{y} \leqq 2$		

Table 30. Comparison between the predicted value of all the hectare fields within the study area (fields below 2,500 m excluded) and the number of Walias which were observed on these fields by H. Hurni in 1975/76. In the contingency table the figures stand for the number of hectare fields. Significance was examined with Chi-square-test

Walia ibex 1975/76, Observations H. Hurni (χ^2: 79.5, df: 8, p $\ll$ 0.01)

Codified number of Walias observed	Prognosis Predicted value for hectare field					
	$\hat{y} \leq 0$	$0 < \hat{y} \leq 0.75$	$0.75 < \hat{y} \leq 1.5$	$1.5 < \hat{y} \leq 2$	$\hat{y} > 2$	total
0	1,893	1,246	949	451	284	4,823
1 or 2	3	9	13	4	8	37
≥ 3	0	6	13	1	9	29
Total	1,896	1,261	975	456	301	4,889

However, in a number of places the given prognosis for the Walia ibex in Fig. 35 cannot correspond to the true pattern. These "errors" are quite evidently caused by a factor that is connected with the space, in some cases presumably with a critical minimum size of suitable terrain, that is certainly larger than only a few hectares; and in other cases possibly with the shape and geographical situation of the home ranges. In a number of small, but steep-sided valleys on the high plateau, the corresponding hectare fields have received high values (all these valleys are located in the southern part of the map along streams flowing into the Jinbar Wenz (River), which runs on the high plateau from east to southwest). Nevertheless we shall not find any ibexes in these places, as they are too small to carry their own population or too isolated from the escarpment to permit a regular exchange within a possible home range. All such hectare fields would represent suitable Walia range only if they were situated near or within the precipice. But we can also find analogous errors in the reverse sense. In some places, for instance on the crest near the peak of Imet Gogo or in two fields within the escarpment range north of Kedadit, we find open circles indicating a negative value for ibexes. Hectare fields in such geographical situations are in fact included in the ibex range, although their local quality for ibexes might be low.

With regard to the location and the size of suitable terrain these "errors" of the prognosis therefore encourage the following general conclusions: (1) Small places of high local quality, but largely isolated within unsuitable terrain are worthless for ibexes. They may be considered extreme cases of small habitat islands (Mac Arthur 1972). (2) Places of low local quality but situated within an extended ibex range will be included in this range and therefore upvalued.

The question arises however, as to how the ibexes respond to fields of different local value that are situated in a larger, generally suitable terrain. It was therefore checked whether or not the field classes were correlated with group size, certain activities of the animals and frequency of particular classes of the Walia ibex. A first comparison showed an obvious correlation with the group size. Considering the five classes of suitability of the fields for the Walia ibex (Fig. 35) the average group size calculated for ibex aggregations observed on fields of these classes was in the sequence from the best class to the worst: 5.1, 3.9, 3.3, 2.9 and 2.4. Thus, with decreasing field quality the average group size decreases. A second comparison, focussing on the animals' activity on the fields of different classes, does not reveal a similarly clear result. However, in comparing the activity classes: standing, lying, feeding and moving, it was seen that the relative frequency of moving in terms of walking or running was lowest in the two best field classes and highest in the (pooled) two poorest field classes. But the differences, particularly between feeding, standing and lying seemed to be of an incidental character. In a contingency table, the frequency of the four distinguished activities was compared with the field classes (of these, the two poorest categories were pooled because of otherwise too low expectancy values). The Chi-square did not reach the level of significance ($\chi^2 = 3.71$, df. 9, n.s.). A third comparison showed that the distinct ibex classes did not frequent the fields of different quality in the same manner. The result of this comparison is shown below in the context of the discussion of the social organization of the Walia ibex (Chap. 16.1, Table 31).

According to the basis given in the habitat maps, calculated and presented for the Walia ibex and the Klipspringer, and after having also discussed the niche differentiation of the mammals included in Chapter 12, the question whether the various herbivorous species have managed to be ideally coadapted arises as a next conclusive step. A balanced utilization of the available plant biomass, i.e., an equally optimal density of foraging animals in the various types of habitat, would be an indication for ideal coadaptation. However, a statistical examination of this question has to be postponed until quantitative data are gathered on the biomass available to the herbivores in each habitat type and each hectare field separately, on habitat preferences and densities of further species such as Duiker and small mammals, and, in a more differentiated way, with consideration of the activity pattern, for the Gelada baboon.

Fig. 36. Theoretical distribution pattern of the Klipspringer within the study area according to the regression equation. As symbols, in the sequence of increasing values of the hectare fields, *open circles* were applied and *small dots* up to *large dots*. The same five classes were used in Table 29 (there, the two highest classes were pooled). The distribution pattern may be considered a habitat map of the Klipspringer

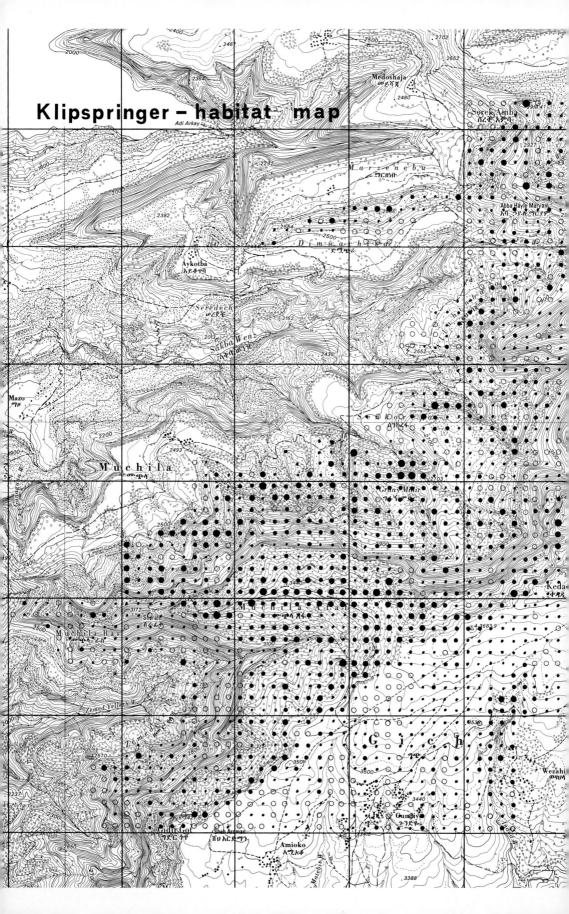

field values

ŷ ≦ 0   0.5 < ŷ ≦ 1   1.5 < ŷ

0 < ŷ ≦ 0.5   1 < ŷ ≦ 1.5

Summary to Chapters 10 to 14

In Chapters 10 to 14 the habitats of the Walia ibex, the Klipspringer, the Gelada baboon and four further mammal species are described by analysing quantitative data. The results are designed to show the differencies between the habitat selection by each species and – for the Walia ibex only – I have tried to show relations to the history of climate. In considering these items the reader is invited to imagine the features in relief and vegetation of this afroalpine area. This can be done by consulting figures 2 to 11 (pp. 8 to 25), and particularly Table 11 (p. 71), in which environmental characteristics at different altitudes are presented.

As expected from a member of the most alpinistic ungulates, the Walia ibex has its main range in the steep, rocky and topographically heterogenous slopes that are mainly found between 2,800 and 3,400 m. The entire range of the Walia ibex, however, extends from roughly 2,500 m up to the edge of the plateau, but excludes the plateau itself. Fresh ground vegetation seems to augment the habitat-value for Walias. This is indicated by the relatively high number of animals observed in the long-grass mountain steppe, in the much frequented generally east-facing slopes that receive more rain than the slopes facing west, in the seasonal changes of the preferred compass directions, and in the positive contribution of the factor troughs – but not ridges – in the multiple regression. In a habitat map (Fig. 35, p.126), the various characteristics of the Walia range are considered and visualized. It is a type of habitat typical for a Capra. However, apart from specific afroalpine characteristics, as one deviation from the habitat of other members, the high amount of forested areas where animals were observed has to be mentioned. This peculiarity has to be seen in the light of the history of climate. During the time the Walia ibex invaded Simen, presumably 12,000 to 20,000 years BP, the timberline was roughly at 2,800 m, and thus main escarpment ranges were above the forests. The actual timberline in natural areas, however, is at 3,600 m and reaches the edge of the plateau. The change may have forced the Walia ibex to adapt to or tolerate forested ranges.

In the natural montane forests of the lower altitudes, the vegetation belt being under highest human pressure, the ranges of the Walia ibex overlap with those of the Bushbuck, Bushpig and the Gureza. The ericaceous and the afroalpine belt is shared by the Walia ibex with the Klipspringer and the Gelada baboon. In contrast to the ibex the Klipspringer ranges more often in areas of moderate gradient of slope, in forests, in topographically exposed areas and, less frequently, in troughs. It is an open question whether the definitely smoother terrain of the Klipspringer, as compared to the ibex habitat, may be considered an effect of a divergence caused by interspecific competition between the two ungulates. The Gelada baboon frequents extremely steep as well as flat terrain and open vegetation. Among the three herbivorous mammal species living in the same altitudinal range and occupying overlapping but obviously different niches, the Gelada baboon is the only one to forage extensively on the flat plateau. It frequents open and exposed habitats but searches for protection when there is strong wind and/or no sun. The calculated coefficients of association and a peaceful interspecific behaviour indicate that there is no interference between the three species. It has been speculated that a mutual benefit in anti-predation might be one key to the understanding of this interspecific situation.

The Social System of the Walia Ibex

15 Towards the Social System of the Walia Ibex: Approach, Inferences and Predictions

15.1 The Approach

In the previous chapters the ecological niche of the Walia ibex within the environmental situation in Simen has been described. Sex and age classes of the ibex have not yet been distinguished. They will now be treated separately. At the same time, comparisons with other species living in Simen will only be made in exceptional cases. With the intention of comprehending more thoroughly the biology and the niche of the Walia ibex, an attempt is undertaken in this second major part of the study to understand the social system of this subspecies of Caprinae. A basic presupposition is that natural selection is a process that optimizes, e.g., the morphological, physiological, bionomic and social nature of an animal – all being interdependent characteristics – in their adaptation to environmental conditions. In so far that this is correct, it must be possible to make preliminary inferences on the social behaviour of the sex-age classes of a population on the basis of the existing information: According to the literature and the questions of research that were given first attention in this study, so far data were available on the morphology of the Walia ibex (e.g., Rüppell 1835; Nievergelt 1972a; Zingg 1980; Chap. 4.1, p. 28), on its reproductive cycle and general grouping pattern (Nievergelt 1970a, 1974; Chap. 4.3, p. 35) and particularly on its habitat preferences (Chaps. 10–14).

This state of knowledge is considered as the framework of conditioning factors within which *inferences* will be made. They are *not* considered conditioning factors of any evolutionary process. The procedure is the following: Chapter 15.2 gives the inferences on the presumably most adaptive social behaviour of the sex-age classes and on differences in their ecological behaviour. In Chapter 15.3 some more detailed predictions on the level of the data are deduced, and in Chapter 16 these predictions are separately examined. This procedure is not circular because the results of these examinations were not consciously known at the time the inferences and the subsequent detailed predictions were made. For most predictions I had in fact no idea about the outcome at that stage. Only in the case of prediction 6, did I feel that the result of the analysis would have to be roughly the way it actually came out.

While reading the inferences on the sex-age classes and the directly testable predictions, one has to consider that these predictions cannot be at a particularly advanced stage, as, during this field study, observations on the social behaviour did not have priority (see Chap. 7.1, p. 53), neither – with few exceptions – were the animals known individually.

Concerning this whole approach we assume this ibex subspecies to be closely adapted to its environment. It is, however, theoretically possible that this assumption is not completely correct. I admit that this could in fact be due to changes of the climate and/or due to earlier adaptations in the evolution of the Caprinae in other geographical regions. However, in view of the 12,000 to 20,000 years of life in this afroalpine region, as was supposed in Chapter 4.2, I assume that optimal adaptation has been reached.

15.2 Inferred Behaviour of Distinct Ibex Classes

Among the facts and evidence to be considered as conditioning factors, some are particularly important. They will be recalled in the following: In the Walia ibex, there is conspicuous dimorphism between males and females; and among males, various age classes are easily distinguishable by morphological characteristics such as body weight, horn length and beard size (see Chap. 4.1, Fig. 14). As in other Caprini, the males are attributed with prominent rank symbols which gradually increase in size (see also Schaller 1977). The reproductive cycle is characterized by rutting throughout the year, but with a definite rutting peak, which is more pronounced in older animals. The observed mean group size was around three animals without considerable changes over the seasons (see Table 7 and Nievergelt 1974). The number of males associated with females decreases with increasing age (see Fig. 17, p. 37). The mating system is supposed to be polygynous, as suggested by the average numbers of males and females in mixed groups (see p. 38 and Fig. 4 of Nievergelt 1974). For the development of characteristics of ibexes such as sexual dimorphism and polygyny, which can be promoted by a heterogenous habitat I refer to Orians (1969), Wilson (1975) and Ricklefs (1979), as quoted on p. 38.

The Simen mountains are distinguished by a winterless afroalpine climate. Irrespective of some changes in the seasonal suitability of the various habitats, the whole range can be used and foraged throughout the year and because of local heterogeneity each geographical unit or even subunit is able to satisfy all the ecological demands of an ibex in all seasons. Of course, this does not mean that all these units are qualitatively equal. Each unit was frequented by all of the distinguished sex-age classes (Table 7, p. 60).

Knowing this much about the Walia and its habitat and considering also the habitat preferences that were shown in Chapters 10 to 14 we now ask the question: what is the optimal behaviour selected by each of the ibex classes, that provides these class members with the maximum chance for their genes to survive and multiply? For ease of separation, the various inferences corresponding to each of the sex-age classes are numbered. I should like to state that these inferences imply simple guesses as well as conclusions. When they were deduced from particular main facts, the latter were mentioned. In the drawing of the inferences I was also affected by my own perception of the ecology of this ibex and this region. Undoubtedly, this was the more the case, the less the inferences were logically necessary. For a description of the sex-age classes see p. 30.

1. Inferred behaviour of old, dominating males: Assuming that these animals are of an average age of roughly 7 to 11 years, we must consider that they have already invested 7 years of foraging, growing and collecting experience until they have reached this favourable position which "guarantees" an open access to females in oestrus (see also Aeschbacher 1978), but which lasts only a few years. Given the small average group size, it is not likely to be optimal behaviour for these adult males to live in closed groups and keep only one female or perhaps two or three. In pair bonds sexual monomorphism is the rule (see Orians 1969; Estes 1974; Jarman 1974). In a monogamous system with an equal sex ratio, a young adult male would already be bonded and would hardly invest in such prominent rank symbols. In the Walia ibex we have to expect therefore that dominating males have developed a behaviour or tactic which permits them to contact various female groups, and take the utmost benefit of the few years of being in a dominating position. The proportion of males above seven years to adult females observed was 85 to 721 (Table 7), and thus the low number of old males seems to permit such a general system. However, it is essential for these males to find their groups regularly and this constraint presupposes that they are informed about the range habits of the female groups being checked. An opportune method of learning the females' habits and consequently being able to predict their sites seems to consist in subordinating themselves within the groups and watching the movements of the members carefully instead of quickly taking over the leadership. Thanks to the prominent rank symbols, there is still no doubt about the actual dominance of the old male and a smooth access, be it to a group as a whole, or to a particular female, cannot be endangered by this active, partial subordination. The suitable and possibly optimal tactic of the old males therefore consists in playing the role of a visitor in the female-juvenile groups and of a follower within these groups, and to include several groups of this kind in their home ranges. Such a tactic of the dominant males with frequent changes from group to group would be backed also by the following two facts: First, males above 7 years are the class where most lone animals were observed (for the male classes see Fig. 17, p. 37, for the comparison with females see Fig. 5 in Nievergelt 1974). Second, in Table 7 (p. 60) the number of animals observed from different observation points is listed. From these data it can be surmised that animals of all sex-age classes occur in each geographical unit or even sub-unit of the area. This would differ for instance from the situation of the Alpine ibex outside the rutting season where we usually have areas with males only and others with females with young (Nievergelt 1966a; Schaerer 1977).

If it is accepted that dominating males will check various female groups we can infer that these males should have larger home ranges than females. This would be a phenomenon such as was described recently in the social system of the badger, where females had restricted borders within the males' range. (Kruuk 1978), and it is possibly similar to the system of the Soay-sheep in rut (Grubb and Jewell 1966).

Apart from the behaviour focussed towards females within the home range of a dominating male, the question arises of whether or not other old males might interfere and in what way a male defends his exclusive access. Territoriality would be one way to achieve an accepted range. In the reasoning about other possibilities to organize competition we may consider the following: In discussing the lack of a breeding territory in sheep Geist (1971a, p. 228) points out: "...breeding territories appear to be typical of ungulates in which adult males look alike, so that the rank (fighting potential) of each cannot be predicted from his horn or body size. The breeding territory can serve in this situation as rank symbol". Adult Walia males, as was written above, as other ibexes and sheep, do not look alike. Independent of a particular place, horn and body size may function as rank symbol. Possible rivals of a dominating male are necessarily of a similar age, and therefore there will always be a small number of potential intruders only. This seems to be promoted by the surmised behaviour of ibexes to remain relatively often in their traditional ranges: Young ibexes, like young sheep are likely to grow up within the area or close neighbourhood of the mother's range and are not forced to leave familiar terrain (Geist 1971a). Due to such characteristics a system without territoriality might be sufficient. Young males, living within the range of a dominating male, (and to which a fairly high degree of genetic relationship is to be assumed), could give additional help and function to space out younger immigrants with a lower degree of genetic relationship.

2. Inferred behaviour of young males; in this context males of roughly 2 to 4 years are included. Corresponding to an ordinary age-distribution pattern, this young male class is fairly well represented. In my observations, males of these ages appear with a frequency of 388 sightings compared with the 721 females (overall sightings: about 3 years and older, Table 7). In fact it is a class which is in a state of transition; in their morphological development they have just passed the horn length and body weight of adult females, but, as yet, there is not much sexual dimorphism, and their behaviour in grouping reflects slightly loosening ties to the female-juvenile groups (Fig. 17, p. 37). Due to this state of transition we have to deal with a social behaviour which is focussed not only on the reproductive success of the moment but also on an optimal preparation of a probable future position. An optimal tactic is most probably to follow frequently distinct female groups. In practising this, the young males have the chance of learning from an experienced adult female how to range in the terrain, and how to respond appropriately to environmental conditions. At the same time they might be able to copulate occasionally, if a dominating male fails to join the group while a female is in oestrus. Similarly age- and rank-dependent mating behaviour as predicted here have been described for the Yellow baboon, *Papio cynocephalus* (Hausfater 1975). I guess, with reference to the momentary situation, that the young males' tactic within the female group has some similarities to the behaviour within the pair-bonds in monogamous species, e.g., in Duikers (Jarman 1974).

However, for the young males, the question of being tolerated in the female-juvenile groups occurs, be it by the females or by the "circulating" old males. First, why should the young males be tolerated by the females? I presume that the interests of the females are conflicting. Clearly, as feeding competitors the young males are a burden in the female groups, but with the physical superiority that will develop or has possibly already developed, the females' possibilities of excluding him are apparently limited. Moreover, if the young male is an earlier kid of the leading female, it could be in the interests of their own genes too to keep the son close. Second, what could, besides economic reasons, explain the supposed tolerance by the old males? By them the young males might be tolerated because, even though they may be sneakers, they subordinate clearly, also "morphologically".

Functionally their presence and occasional courting might stimulate the physiological and behavioural readiness of the females to rut. In addition it may be possible that the courting and rutting behaviour of the young males helps to signalize from afar the presence of a female being in oestrus to the dominating male (a signalizing effect to other males was presumed by Aeschbacher 1978 for the Alpine ibex).

With reference to the future position, success in social competition with other males of similar age has to be achieved. Thus practising fighting is essential and this need automatically promotes the formation of young-male groups. With increasing age this aspect becomes more important; statistical evidence for this tendency is seen in Fig. 17.

3. Inferred behaviour of medium-age males of roughly 4 to 7 years: This is the class between the two discussed above. These males are at the age when sexual dimorphism actually develops. They can be considered as in the proximate waiting position for dominance and therefore the pursuit of establishing rank among other males most likely dictates the optimal tactic. This inference is backed by the information given in Fig. 17 (p. 37). In this figure it can be seen that medium-age males were relatively more often associated with other males than are any other age classes of males. In the Alpine ibex it was shown that there is the most tension between the dominating male of a group and the male ranking next (Nievergelt 1967; Aeschbacher 1978). For the medium-age males it is therefore assumed that remaining attached to female groups could become more difficult in the occasional presence of an older and stronger male. As was shown in Fig. 17, contacts to the female-juvenile groups are not stopped, but compared with younger males, visits in these groups are more scarce and/or less extended.

The age of change from the behaviour of the medium-aged males to that of the dominating males was indicated to be 7 years (see p. 30). In fact, this age cannot be considered as generally valid, as the precise age depends on various factors such as the constitution of the individual and the age structure within the population. If the dominating class is well represented, it may well be that males of 8 or 9 years are still acting according to the postulated behaviour of the medium-aged males.

According to the inferred behaviour of the various age classes of males, I assume that – on the average – the reproductive period of age is shorter and begins later as compared to this period in the females. However, as yet I have no evidence for a physiologically determined sexual bimaturism (see Wilson 1975).

4. Inferred behaviour of females: The pronounced sexual dimorphism in the ibex, with females half the weight of adult males, and in addition, the costs of pregnancy and lactation should raise the energetic requirements (Kleiber 1961; Geist 1974a, b; Schaller 1977). According to measurements in other ungulates, we may assume a relatively small acceleration in these requirements during pregnancy, but a sharp increase to around double the value for roughly 2 months during lactation (Moen 1973; Ellenberg 1978; or 1 month Weiner 1977). According to laboratory experiments the costs of pregnancy and lactation in most small mammals demand an increase in the required energy of at least 60% for each pregnancy (Kaczmarski 1966; Campbell 1974). Due to these considerations on body size and costs of reproduction the per-gram metabolic rate in females is supposed to be higher than in males. Therefore, females ought to be more careful in

choosing food of a high nutritional content, and they should select their ranges more strictly according to optimal foraging conditions.

In consequence of the inferred tactic of the dominating males, females should use relatively less extended ranges than males. If this were true, we have to ask the question, in which way the behaviour of limiting the home range size would correspond to other inferred characteristics of the females. Apparently, it does not seem to jeopardize the above inferred need for high-quality food: Ranging in the afroalpine and heterogenous area of the Simen mountains, even within relatively small areas, an animal can find such food all through the year (see p. 24). By remaining in a relatively limited area, a female gets to know precisely the selected home range, and this knowledge may favour her in avoiding predators and in escaping from them. Predators with their much larger ranges presumably do not have a chance to accumulate the same experience about each local area. A limitation in the home ranges of each particular female (not a limitation of the area used by all of the females) most likely is also an adaptation for the security of the offspring, particularly of the type such as goat kids which follow their mother shortly after parturition, running around and playing. We can assume that the kids become familiar with their environment more quickly, the more limited the range being frequented by their mothers. By this method of limiting the home-range size we may also assume that the dominant males are able to find the various female groups more easily. This facilitation for the males must be also in the females' interest, as "errors in mate selection are more serious for females than for males" (Orians 1969, p. 591).

From the point of view of a possible difference in the home-range sizes the behaviour of old Walia males seems more risky. In this rocky and often inaccessible terrain a larger home range most likely requires more time to explore it in order to reach a state of familiarity with the environment that guarantees a suitable anti-predation behaviour. With the smaller home range inferred for females, we may assume that adult females have a more detailed knowledge of their environment, including suitable feeding places, hides and inconspicuous paths.

Given the above inferred characteristics of the females with the comparatively higher demands in the required energy and the suggested smaller home ranges, the inference is put forward that large all-female groups and female-juvenile groups ought to be avoided, so that the local feeding sites are in less danger of being over-grazed (for theoretical considerations on the relations between feeding style, body size and group size in antilopes see Jarman 1974). A possible mechanism in females to keep their groups small is the observed seasonal intolerance of females against other females. This intolerance is abandoned during parturition, which could be an adaptation to the omnipresence of birds of prey. Data and discussion of this temporary intolerance were given in Nievergelt (1974, p. 339). Due to the small population size and density, at present little pressure seems to be needed to keep group sizes in female-juvenile associations low. However, there is some evidence for a regulating mechanism to act at high population density. This evidence originates from observations in areas with large populations (and thus a presumably high population density) of the Alpine ibex, *Capra i. ibex* in Switzerland and the Wild goat *Capra aegagrus* in Iran. Both species show a similar sexual dimorphism with definitely smaller females, as with the Walia ibex. The nutritional requirements may therefore be comparable. There seems to be a tendency in females to use smaller home ranges than males also in these two species (see p. 168). According to my observations, large associations above approximately 50 animals were, in fact, exclusively all-male groups.

5. Inferred behaviour of kids: In this class are included newly born animals up to the age of roughly one year. They should aim to stay as close as possible to their mother, who provides protection – particularly against birds of prey – and guidance, and they should preserve this bond as long and as exclusively as possible. This behaviour would also include the preventing of copulations between any male and their mother.

6. Inferred behaviour of young animals of 1–2 years: This is another class in a typical state of transition. They can no longer be considered kids, but they are still morphologically and physiologically immature. Their optimal tactic seems to be to remain in a female- or female-juvenile group, preferably in the group with their mother, for as long as they are tolerated. The association in such a group provides the combined possibility of getting more experience and being protected, thus being at least loosely under a skilled leadership.

15.3 Predictions Deduced from the Inferred Behaviour of the Sex-Age Classes

The described inferred behaviour of the six distinct classes of the Walia ibex is based on the results reported in previous chapters. Additionally, some distinctive characteristics of each of the classes are implemented as theoretical requirements. In the following, these requirements are presented as a list of detailed predictions.

1. Viewing the inferred tactics of the males of various classes and those of the females we may assume that the behaviour of the latter tends to be more ecologically governed, whereas the movements of the males seem to be mainly ordered according to social conditions. On the basis of this assumption I put forward the prediction that the habitat pattern of the males compared with the females is closer to the expectancy pattern arising where there is no habitat preference, and that the habitat frequented by the females is more precisely determined as to local ecological conditions than the expected wider habitat (or niche) of the males. For the same reasons we may predict that in the theoretical distribution pattern as given in Chapter 14, the relative frequency of animals using fields of low quality is higher in adult males and vice versa.

2. It was inferred that the optimal tactic of young males is to stick to particular female groups more consistently than old males, which are inferred to visit female groups only briefly. We have therefore to expect that the habitat pattern of the males gradually deviates from that of the females as they get older.

3. This prediction is based on the inference that adult males visit the females and not vice versa. As in other Caprinae, it is probable that males and females have different general habitat preferences (Nievergelt 1966a; Geist 1971a, 1977), although these preferences for the males may be less pronounced (see the first prediction). As a further presupposition we may assume that, due to the low

population density in the area, it is most likely that any class of ibexes will find the places it prefers, regardless of whether it is dominating or not. One can deduce from all this that the expected habitat difference between males in all-male groups and males in mixed groups is greater than between females in all-female groups and females in mixed groups. The compromise in the habitats frequented by males and females in mixed groups is predicted to be smaller for females than for males.

4. According to the inference that dominating males will check various all-female and female-juvenile groups, it is predicted that old males will tend to have larger home ranges than females. Viewing the inferred behaviour of old males and of females I guess that the possibly smaller size of the home ranges of the females is not likely to be the result of an unknown behavioural effort of the males. I assume, however, that the females' ranges are mainly determined by habitat conditions. It seems possible that the average size may be increased within a relatively poorer area. It could also increase where there is a more pronounced tendency to use different parts of the area in different seasons. Theoretically it could also be possible and still coincide with the old-males' behaviour described above, that each of several female groups use the whole and the same home range as the old male. This seems unlikely, particularly in considering the inferred behaviour of the females. However, because this more theoretical case is possible, the above prediction is described as a tendency.

5. Assuming that the tactic of old males is to visit regularly the various female groups, we may expect some indications that adult males join or leave groups while we were observing them. Generally we might expect the males to be relatively loosely associated with the female-juvenile groups.

6. In order to be successful, the tactic of old males implies a rank order that is immediately respected between males of different classes. In the Alpine ibex it was seen that in cases of clear age differences between two rivals no fight, or almost none, was recorded. By analogy, in the Walia ibex no fights or very few are expected between males of different and well-attributed age classes. Fights are predicted to take place in cases of subtle rank differences between males, but also between young males and females in the phase of rank change. Furthermore we may put forward the possibility that males fight relatively more often in the phase where there is a need to establish rank, and thus before having reached their dominating position rather than afterwards. The rank position in the rutting situation may appear to change between male and female, when adult males are actively subordinating themselves.

7. Adult males were inferred not to lead when joining females and young; in contrast, they were expected to watch carefully from behind and approach only the rutting females. The prediction is therefore offered that in the marching order, adult females are likely to assume the leading position. For young and adult males, an average position at the back is expected.

8. Associated with the kid's "interest" in preserving a bond with its mother, it was expected to undertake behavioural efforts aimed at preventing copulations between its mother and any approaching males. However, no indication of such a behaviour could be observed. The location of the kids in the mixed groups in relation to their mother's position and that of a rutting male did not reveal separating efforts from the kid's side (in contrast to all the other predictions, that

will be treated in the next chapter, the outcome of this examination with its scant character is given here).

9. Females were supposed to require more energy and protein per unit of body-weight per day than the larger males. An even higher relative metabolic rate may be assumed for the Klipspringer (Bell 1970, 1971; Geist 1974b). It is predicted that this inferred difference is accompanied by a difference in food selection. In a rough comparison between Walia males and females and the Klipspringers, the food taken is expected to decrease in its absolute quantity, but to increase in quality and/or the degree to which it must be utilized.

Despite this framework of metabolic background, it is not yet possible to predict in a conclusive manner detailed feeding tactics of the sex-age classes, as basic information is missing. Such information as, for example, the actual required energy of Walia males and females, and of the Klipspringer, the capacity, efficiency and physiology of the digestive apparatus, including mouth and teeth, as well as the nutritional content of the available plants in different parts and phases is not available. The subsequent prediction is therefore based on slight evidence only.

Each herbivorous animal has the choice of selecting between different plant species to maintain diversity of diet. It can also select according to a specific flowering phase and/or certain especially digestible parts of a plant. High quality foods are definitely the fruits of the plants. However, in the phase of ripe fruits, the stems and leaves are often dry and of poor value, so that each plant is able to contribute only small portions of valuable food. The average nutrient content of the whole plant is possibly higher when freshly grown. Thus, if much food is required, the better tactic could be to select young, green plants instead of searching for seeds. We may therefore predict that animals requiring comparatively small amounts of food, but of better quality – as the females of the Walias in comparison with the males – may show a tendency to forage plants in the fruit phase.

In the approach presented in this chapter, originating from a number of facts considered as conditioning factors, inferences on the adaptive social behaviour of the sex-age classes of the Walia ibex have been described (Chap. 15.2). Subsequently a number of detailed predictions were deduced (Chap. 15.3). It is expected that their analysis given below will provide a suitable basis to judge the plausibility of the given inferences, and thus will lead to an improved understanding of the social and ecological behaviour of the Walia ibex. In the following chapter, the above predictions 1 to 7 and 9 will be tested and discussed. For the most part, comparisons between the classes, mainly between males and females, are carried out. However, in Chapter 16.6 on plant selection, the Klipspringer is also included. Obviously, it is not possible to treat all these predictions fully or to give satisfactory answers to all the questions. The investigation of a number of questions would require large numbers of individually recognizable animals and/or possibility of long-term observations. Such questions cannot be answered yet, and must be treated on the level of qualitative impressions formed from certain observations.

16 The Social System of the Walia Ibex; Evidence for and Tests of the Above Predictions

16.1 Habitat Preference of Males and Females of the Walia Ibex; Tests of the Predictions 1 to 3

The first three predictions are treated jointly in this section, as all of them deal with habitat preferences of the animals, and several items in the data presented provide information on more than one of these predictions.

Table 31 is linked with the question on the response of the Walia ibex classes to hectare fields that are of a different value according to the habitat map (Chap. 14, Fig. 35 and p. 126), and also with prediction 1 stating that females tend to be more ecologically governed, whereas the behaviour of males is performed mainly according to social criteria. As was predicted, adult males are represented most on fields of low value, and least on fields with the better habitat conditions. The "rank order" is based on the relative frequency with which these fields were visited by the different Walia classes. Females seem to avoid fields of poor habitat value, and they have a higher frequency on better quality fields than adult males. The highest frequencies in the columns that represent fields of the highest value were reached by the kid class. As kids were always spotted together with females, the class of the kids may also be considered as a class of mothers with their offspring. Considering also the above-mentioned costly period of lactation (see p. 141), the concentration of this class on the most valuable fields appears almost as a logically necessary result.

Table 31. The rank order of the relative frequency of five Walia ibex classes to visit hectare fields of the five different values according to the prognosis as shown in Fig. 35. In each column the rank order is given for a field-value level. Field values are based on the stepwise multiple regression that has led to a prognosis of the Walia habitat. Fields with the best habitat conditions are on the right side, those with the poorest habitat on the left

Rank in relative frequency	Predicted value for hectare field				
	$\hat{y} \leq 0$	$0 < \hat{y} \leq 0.75$	$0.75 < \hat{y} \leq 1.5$	$1.5 < \hat{y} \leq 2$	$\hat{y} > 2$
1	Males >4	Males >4	Females	Kids	Kids
2	Young	Males ≤4	Males >4	Young	Young
3	Males ≤4	Kids	Young	Females	Males ≤4
4	Kids	Females	Males ≤4	Males ≤4	Females
5	Females	Young	Kids	Males >4	Males >4

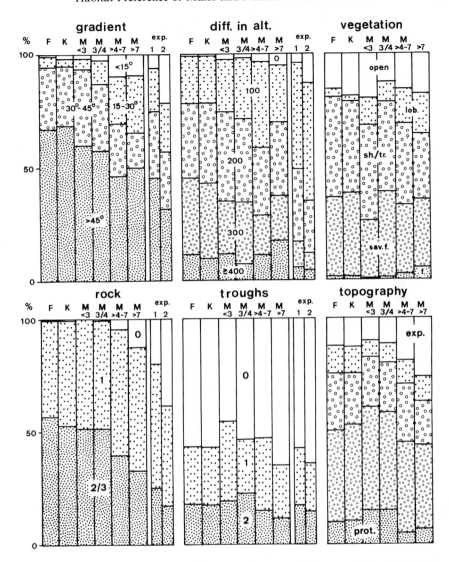

Fig. 37. The frequency pattern for six different environmental factors and for six Walia ibex classes. The following four field and two observation factors were considered: the gradient of the slope (see Table 10), the amount of rock coverage (Table 12), the difference in altitude to neighbouring field in metres (Table 12, therein combined with the gradient), the quality and number of troughs (Table 12), the type of vegetation (Fig. 29; open vegetation in the form of mountain steppe, open with Lobelias, single shrubs or trees, savanna-type-forest, dense forest) and the topography (six classes in a gradient from exposed to protected, see also Tables 20, 23). Columns within each of these factors stand for females (*F*), kids (*K*), and males of four age classes (*M* < 3, *M* 3/4, *M* > 4–7, *M* > 7). For the field factors two expectancy patterns for random distribution are given: *Exp. 1* represents the fields within the study area with a degree of human influence of the classes 0 and 1 only (Tables 10, 12); *Exp. 2* represents all the fields of the study area. The given frequencies are based on the following animal numbers: *F* 646–700; *K* 297–335; *M* < 3 160–174; *M* 3/4 181–199; *M* > 4–7 164–186; *M* > 7 73–80; due to some incomplete observations the animal numbers have to be given with a range, the *lower numbers* referring to the observation factors

Thus, the whole pattern shown in Table 31 can be considered consistent with prediction 1 (p. 143).

A similar effect is shown in Fig. 37, but these data are based on a different approach. The figure is composed of six sub-units, each representing a different environmental factor. Therein the data are presented in columns for the Walia ibex classes distinguished. The four environmental factors to the left are field factors, and therefore an expectation of random distribution could be added. The two factors to the right without random expectation are observation factors (see p. 67, Chap. 8.1). Within each column and with different shading from white to dark grey I have shown the relative frequency covered by the different levels of the factors. In accordance with prediction 1, in a comparison between the Walia classes and the random expectation, we can see that the pattern for adult males approaches most closely towards the random distribution, although with a different degree of evidence in the four field factors. Prediction 2 can be examined without reference to a random distribution pattern, and so all six factors can be considered. It was stated that the habitat pattern of the males should gradually deviate from that of the females with increasing age. The data shown in the figure are consistent with this prediction.

Figure 38 presents data for all three predictions to be discussed in this section. In triangle coordinates the factors gradient (above) and type of vegetation (below) are dealt with i.e., the same two variables that were used in Tables 24 and 25 (pp. 104 and 112). In these diagrams on the left the average values for each quarter of the year are shown separately for males and females in single-sex groups. The corresponding points for the four seasons were connected for each of the sexes. The surface covered may be used as a rough estimate of the ecological range for this factor in males and females, utilized over the year. For both factors the range of the males seems wider and less extreme (prediction 1). The result, however, is not fully conclusive, as the "ecological range" shown is based on differences between seasonal averages only. But it indicates that females show less extreme habitat changes, at least with regard to the gradient and the type of vegetation. A number of further environmental factors, not shown here, were examined in the same way. In most cases a similar result was observed. As a general rule males showed a tendency towards less extreme habitats than females. In comparing the surfaces in the diagrams that indicate the ecological ranges of males and females, in a few exceptions only those of the females reached a similar size as those of the males. In the diagrams to the left, from each corner point, a line was drawn. These straight lines link only the points referring to single-sex groups with the points of the same sex class and season, but when associated in mixed groups. The length of this line may be considered an estimate of the "ecological compromise" of a sex class, when joining mixed groups. In accordance with prediction 3, the result shows that the compromise seems to be in fact larger for males than for females. Theoretically – one might reason – the compromise ought to be zero for females, as the males were inferred to visit the females and not vice versa. With reference to Fig. 38 this would be an unrealistic expectation. It would ignore, apart from the deviation, unbalanced factors that might influence results such as group size, activity and season. The estimated small compromise of the females, as being visualized in the figure, also touches a further question. Do the females actually remain entirely passive in the forming of mixed groups? This question will be discussed briefly in Chapter 16.7.

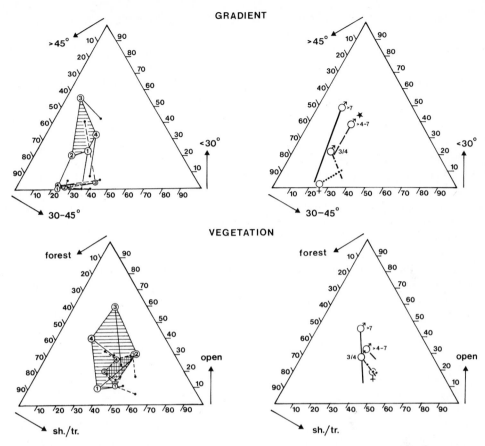

Fig. 38. Diagram in triangle coordinates showing the average habitat selection separately for males and females in single-sex groups and in mixed groups. In the two diagrams *above*, data for the environmental factor gradient of the slope are used, in those *below*, the type of the vegetation. In the diagrams on the *left*, for each of the sexes, the pattern is shown for four seasons: December–February (*1*), March–May (*2*), June–August (*3*) and September–November (*4*), see also Nievergelt (1974). The *shaded surfaces* for males (*hatched horizontally*) and females (*hatched vertically*) represent the pattern in single-sex groups. The *lines from the corner points* connect to the resp. points in mixed groups. In the diagrams on the *right*, the yearly average is given but distinguished for males >7, >4–7 and 3 and 4 years. From the *points* indicating the average in single-sex groups in an analogous manner, the *length of the lines* indicates the discrepancy from the average points of the class in mixed groups. For the gradient only the expectancy value based on the 5,815 fields is drawn with an *asterisk*

In the two diagrams on the right the males are divided into three classes, but the seasons were pooled. Thus we have for each class the yearly average in different single-sex groups, and the connecting lines to the average value of the corresponding class that are in mixed groups. These figures indicate, as might be expected, the widest discrepancy for the dominating male class (prediction 3), but they also show that the males, as they get older, gradually separate from the average habitat of the females (prediction 2).

16.2 Size of Home Ranges; Test of Prediction 4

In Chapter 7.1 it was stated that due to the endangered status of the Walia ibex, no effort was made to trap and mark animals. However, seven Walias were known individually, six by the shape of their horns with respect to the form and sequence of the knots (e.g., Figs. 14a, 15, 39a), and the youngest male by a mark of white colour (Fig. 39b, this mark was applied by M. Demment with a wide needle-less plastic syringe filled with colour fixed at the tip of an arrow). Three of the six older males were individually recognized in an analysis of the photographs taken. The seven individuals are listed in Table 32. Of course, this sample is very small and is only suitable to show some first tendencies with regard to the prediction that old males are likely to occupy larger ranges than do females and young. Additional care is needed, as only four of these animals were recognized on four or more different days. According to the plotted observations, a wide range was seen for animals 1 and 5. I suppose that these ranges cover in reality the whole geographical unit. The animals 2–4 and 6 were always seen within a relatively small area. Animal 7 is a special case, seen first in the Inatye region, thus near Chennek, but appearing in the Kedadit area towards the end of our field period; it was not seen in the Saha unit in between (see Fig. 1, p. 7 and Fig. 23, p. 56). This observation, together with the remark given on animal 1 may be considered an indication – although of weak reliability – of a spacing-out mechanism. The data shown in the table would be consistent with the prediction that older males tend to have large ranges. If the

Fig. 39. Two individually marked ibexes. **a** an old female, animal 4 in Table 32, was naturally marked by a broken horn, **b** the young male, animal 3 in Table 32, was artificially marked with colour (see text). The mark is behind the right shoulder of the animal in front. The animal following is a young female between 1 and 2 years

Table 32. The seven individually known animals with the geographical unit, the sex and age, the number of how often the animals were observed and on how many different days and the resulting range. For the estimation of the size, the observations were plotted on the map. The outermost points were connected and the area including all the points measured

Number of the animal	Animal shown in fig.	Geographical unit	Animal class	Number of observations	Number of days with observations	Resulting size of range	Remarks
1	14a	Saha	Male above 10 years	17	12	$> 3 \text{ km}^2$	No other male of this age, and hornsize was spotted in Saha unit in 1968/69
2		Saha	Male of appr. 8 years	5	3	$< 0.5 \text{ km}^2$	In 3 of the 5 observations it was in the same group as the old male above (3 different days only)
3	39b	Saha	Male of 1 year	6	5	$< 0.5 \text{ km}^2$	
4	39a	Saha	Old female	5	4	$< 0.5 \text{ km}^2$	
5	15	Kedadit: also western edge of Saha	Male of 5 (6) years	6	4	$> 3 \text{ km}^2$	In Kedadit Muchila Afaf-unit very few males of the oldest class were to be seen (see Table 7)
6		Kedadit	Male of 4 years	4	2	$< 0.5 \text{ km}^2$	
7		Chennek Kedadit	Male around 10 years	2	2	(>10 km walk between the 2 points)	First observation in Inatye/Chennek unit east of Saha, second observation in Kedadit unit west of Saha

ranges of the animals, listed in Table 32, are examined separately for each geographical unit, one could surmise a tendency towards the phenomenon that the oldest male of the unit covers comparatively the largest range.

16.3 The Dominant Males Only as Occasional Visitors in Female-Young Groups; Test of Prediction 5

According to the inferred tactic of the dominant male that visits successively the various female-young groups within its home range, some evidence was predicted about males joining or leaving the group, or being only loosely attached to it.

In order to investigate this prediction satisfactorily, long-term observations of certain focal groups and/or a considerable number of individually known animals would be required. Such data are not available.

However, the following two observations might give some indication. They are consistent with the prediction. 1.5.1968; 8.05 a.m.: "a mixed group of 11 animals ran some 100 m when a man passed nearby singing. At 9.00 a.m. the animals were quiet again, and the group had split up: three groups of five, three and two animals respectively were feeding mostly within the same range; definitely separated and 40 m below, within a gorge, was the eleventh animal, a 7- to 8-year old male, the oldest of the group". 20.4.1968; 9.15 a.m.: "at a distance of 100 m above each other there were two very similar mixed groups: a female, kid, yearling, with a male of 5 years, and a female, kid, yearling, with a male of 4 years. Both males showed rutting behaviour. After 30 min, suddenly a 9-year-old male appeared. Its appearance caused some nearby Geladas to run away, and it then passed the upper group where the 5 year old male gave way immediately".

I could not observe single animals, other than adult males, that were apparently joining or leaving a mixed group.

16.4 The Rank Order and the Fights Between Walias of Different Classes; Test of Prediction 6

For the Walia as true ibex with prominent and gradually growing rank symbols, the following was predicted: Inferior group members pay immediate respect to the clearly dominant male and fights occur mainly between animals of the same sex-age-classes. In the following these two aspects are treated successively.

In six cases it was observed that an adult male approached a female in rutting position, while a younger male was standing between the two animals (three times the adult male was older than 7 years, and three times it belonged to the class 5 to 7 years). In each case the response of the younger, clearly inferior male was the same: on the approach of the dominating male it moved immediately to the side; in five cases, and each time the adult male was over 7 years of age, not even a threat on the part of the older male could be seen. Only once, and it was a situation where the estimated difference in age between the two males was a mere 2 years, the dominant male while passing by knocked its horns ostentatiously on the saddle of the inferior

Table 33. The number of observed fights and playfights between animals of the same class and of different classes. Partner A is listed in the rows, partner B in the columns

Partner A	Partner B			Males, ages in years							
	Females	Kids up to yearlings	Undefined young	1	2	3	4	5	6	7	>7
Males, >7										1	
ages in 7									2		
years 6							1	1	1		
5						3	4	3			
4				1		1	3				
3	2	1			3	5					
2	5	2		1	1						
1	5			1							
Undefined young	1		2								
Kids up to yearlings	2	12									
Females	7										

rival. However, all observations support the prediction that fully grown and clearly dominant males do not have to fight for their rank, when entering a group and controlling a female. A similar tendency for absolute respect of rank in ibexes was also described by Schaller (1977), and this coincides too with the Alpine ibex (Nievergelt 1967; Aeschbacher 1978) and the Sibirian ibex (Carlstead 1973).

The number of fights amongst Walias within and between the classes may be considered, with some restrictions, as measure of the effort of the class members to determine rank position. In Table 33 the number of fights and playfights that I could observe between animals of different class combinations is listed. The class of partner A is listed in rows, that of partner B is given in columns. As the table shows, roughly half the fights were observed between animals of the same class. For the males, as was seen in the Alpine ibex, fights occur mainly when the difference in age between the two partners is small (Nievergelt 1967; Schaerer 1977). The figures in the first column, in which the fights are listed, are remarkable when one of the partners was a female. As expected, no fights were observed between females and adult males; but ten when the male was 1 or 2 years old, and thus apparently at the age of rank change. The longest-lasting fight, and definitely the most serious quarrel seen by the observer, occurred on the 14.12.1968. The antagonists were a female and a 2-year-old male, that obviously won.

In order to estimate the readiness to fight within the observed aggregations, the recorded number of fights for a class combination must be compared with the relative time that members of these classes were observed to stay in the same group. This comparison is given in Table 34. For this comparison the ordinary pooled male classes had to be applied and the heterogeneous class of undefined young was ignored. Apart from the results discussed already in the table the following can be

Table 34. The number of observed fights and playfights between animals of the same class and of different classes (obs.) in comparison with expectancy (exp.). The expectancy was calculated on the assumption that each animal within the observed group would fight in equal probability with any other member of the group

Partner A	Partner B	Females	Kids up to yearlings	Males <3	Males 3/4	Males >4–7	Males >7
Males >7	**obs.**	–	–	–	–	**1**	–
	exp.	0.8	0.3	0.1	0.3	0.7	0.4
Males >4–7	**obs.**	–	–	–	**8**	**7**	
	exp.	2.8	1.6	0.8	1.4	1.5	
Males 3/4	**obs.**	**2**	**1**	**4**	**9**		
	exp.	3.8	2.3	1.5	1.3		
Males <3	**obs.**	**10**	**2**	**3**			
	exp.	4.6	3.3	1.5			
Kids up to yearlings	**obs.**	**2**	**12**				
	exp.	15.6	8.8				
Females	**obs.**	**7**					
	exp.	14.5					

seen: The only class that the females fight with more frequently than would be expected in the case of no-class preference are the males below three years. The seven fights observed among females themselves are clearly below the level of expectation. For the fights among males, there seems to be a tendency that old males fight relatively less than young and medium-aged males, which in fact seem to be more involved in efforts to establish their rank. This tendency, which is also shown in Table 33 coincides largely with Schaerer (1977) who found in the Alpine ibex a negative correlation between the age of the males and the daily amount of time spent fighting. As a whole the prediction is confirmed by the data shown in the two tables, that fights occur mainly between animals of subtle rank differences.

16.5 The Marching Order Within the Mixed Groups; Test of Prediction 7

In order to test the prediction that females are supposed to lead mixed groups – a phenomenon that results from the inferred tactic of the dominant males to check the habits of the females – the position of each class in groups of walking or running Walias was examined.

The marching order of these groups was analysed in two ways. In a first approach each animal was given a value between 1 and − 1 according to its position. 1 was given to the leading animal, − 1 to the animal coming last (for this approach see Krämer 1969). Depending on the group size, the values for the intermediate positions changed. For two animals only, the values given were 1 and − 1; for three

Table 35. The average position of Walias of four different classes in mixed groups, when the animals were running or walking. In each group considered the animals were valued according to position from +1 (the leading animal) to −1 (the last animal). Separately for groups of up to five animals and for larger groups the arithmetic mean of all the observed positions ($\bar{x}$), the number of values considered (n), and the standard deviation (s) were given. In the last column the F value of the analysis of variance is given, the degrees of freedom (df), and the level of significance

Group size	$\bar{x}$, n, s	Females	Kids	Young of 1–2 years	Males as from 3 y	F, df, sig.
≦5	$\bar{x}$	0.64	− 0.03	− 0.25	− 0.33	F:11.32
	n	26	12	20	34	df:3/88
	s	0.62	0.53	0.75	0.72	P<0.01
>5	$\bar{x}$	0.24	0.19	− 0.12	− 0.23	F:3.96
	n	29	17	37	34	df:3/113
	s	0.69	0.67	0.56	0.60	P≈0.01

animals: 1, 0, − 1; for four animals: 1, 0.33, −0.33, − 1 etc. Out of all the moving groups the arithmetic mean for each class was calculated separately for mixed groups of up to five animals and for groups above five. The result is given in Table 35. For both group sizes the same order of positions resulted: females – kids – juveniles of 1 to 2 years – males. This order is consistent with the prediction. Females lead, the dominant males are likely to "take" a position behind. Confirmed by a one-way analysis of variance, the differences in the average position of the classes are not likely to be accidental; the F-values are given in the last column. The table also shows that the result is less clear in larger groups. This indicates in larger groups for class members a stronger tendency to be dispersed, or attached to other classes, rather than to be concentrated. The drop in the value for females and the increase for kids when larger groups are compared with smaller groups is apparently caused by a pattern that was frequently observed: the leading female was accompanied by a kid, whereas females following behind were often without a kid, and were usually younger.

This field impression was seen as a sign that the age *within* the sex-classes might be manifested in the marching order. At first this item was examined in females. The same approach as that used in Table 35, but applied to three age classes of females only, revealed the following average positions: old females 0.55 (n:56), medium-aged females 0.31 (n:32), young females −0.41 (n:33). In this comparison, single-sex groups were also considered, but the sequence of the average positions was the same in mixed groups and single-sex groups, when treated separately. Thus, as supposed, within the female class only, the position in the moving groups seems to be correlated with age. As shown in Nievergelt (1972a), there is also a relationship of the age to the fertility rate. The proportion of observations of females with newly-born kids (and of pregnant females) to the number of females being closely followed by rutting males (thus being presumably in oestrus) was for old females 18/11, for medium-aged females 12/22 and for young females 1/26. A very rough estimate of

Table 36. Observed frequence of leading a walking or running mixed group for seven distinct Walia classes compared with the expected frequency if each class is given the same probability of leading a group

Class Walia ibex		Observed frequency of leading a group	Expected frequency for leading
Females		30	13.2
Kids		–	4.3
Young	(1, 2 years)	–	3.2
Males	< 3 years	–	2.4
Males	3/4 years	5	6.6
Males	> 4–7 years	–	3.2
Males	> 7 years	–	2.1
Total		35	35

the ages for the three female classes were, above 5 years for old females, 4 to 5 years for medium-aged females and below 4 years for young females. This relationship and the determined marching order within the females only confirm the field impression given above, that the females with kids were mostly in the leading position.

For the males that were examined secondly, it was noted if in a dyad of males the older had its position before the younger male or vice versa: In single-sex groups the older male was before the younger in 27 cases, but behind in 6 cases. In mixed groups the same figures were 16 and 11. Thus, as a general rule the older, dominant male is usually before its younger companion, but exceptions are common, particularly in mixed groups. According to a four-fields test there is a tendency for the given relations to be different in the two group types (χ^2:3.72, df:1, P $\approx$ 0.06). This difference – if it did not occur accidentally – may be explained in the following way. In mixed groups the adult male takes its position in orientation towards the female(s) and almost ignores the younger male. The position between male classes is therefore less strictly maintained. In all-male groups, however, there is no diverting effect from the females.

In a second approach on the question of the marching order within mixed groups, I observed how often a member of each class was leading a group compared with the probability of equally shared leadership between the classes. The result, given in Table 36, very clearly shows that in most cases females were observed in front of a moving group. The only other class observed in the leading position – but to a definitely minor degree – were young males of 3 to 4 years. It is a class which very often joins female-young groups, but seems to remain loosely associated. I presume that the observed five cases were probably more a case of accidental than of true leadership. As a conclusion, the results shown in Tables 35 and 36 confirm the prediction that very regularly the females and not the dominant adult males take the lead within mixed groups.

16.6 Food Selection of the Walia Ibex of Different Classes and of the Klipspringer with Considerations on Prediction 9

As the title indicates, this chapter does not concentrate exclusively on investigating prediction 9. Data are presented on food selection in the Walia ibex and the Klipspringer in a broader sense, and, stemming from this, emphasis is given to those aspects which are relevant to the prediction. However, two restrictions that reduce the validity of the results must be mentioned. (1) In the feeding observations the various types of vegetation are not considered according to the frequency with which they were utilized by the two ungulate species. The afroalpine mountain steppe is clearly over-represented, and the montane belt, the lowest vegetation zone within the Walia range, is not considered at all. However I assume that existing feeding tactics of the Walia ibex classes and the Klipspringer will be visible in the samples of vegetation units considered. (2) No analyses of the nutritive content of the plant species selected or avoided were carried out. The two techniques applied to investigate the feeding behaviour were described in Chapter 7.5 (p. 65).

In Table 37 the feeding observations which were carried out at a distance are presented. For adult males, for young males and for females the numbers are given for how often members of these classes were seen to forage at particular plant species or at a particular group of plants. Quite clearly, with this distance technique, the numbers of trees and shrubs which are easily visible are over-represented. In most cases, when the animals were feeding in the ground layer, I was not able to see whether they were foraging grasses or herbs. On rare occasions only I could identify the grass or herb species (Fig. 40). From the table, the following can be seen: The Three heather *Erica arborea* is quite regularly foraged by all Walia classes, although – as will be shown below – not intensively. Of the three classes, the young males show the least tendency to browse. In comparing adult males and females, the comparison that affects the test of prediction 9, the preference for the leaves of Lobelias by the males (Fig. 40) and for *Helichrysum citrispinum* by the females is

Table 37. Feeding observations at a distance. For two male classes and for females of the Walia ibex it is listed how often animals of these classes were seen foraging at the plant species or groups of plants listed (see also Fig. 40)

Plant species or groups of plants and water	Males of 3 or more years	Younger males	Females and kids
Erica arborea	26	5	17
Hypericum revolutum	6	–	3
Lobelia rhynchopetalum			
yellow leaves	18	3	4
green leaves	5	–	–
Helichrysum citrispinum	10	12	25
various shrubs	19	1	14
various grasses	8	5	10
various herbs	3	2	9
lichens (mainly *Usnea* sp.)	9	1	3
drinking of water	1	–	1

Fig. 40. In **a, c, e** an old, roughly 9- to 10-year-old male is shown foraging leaves of Giant Lobelia (**a**), herbs in between tussocks of mainly *Festuca macrophylla* (**c**), and *Erica arborea* (**e**). The three figures show that while observing the ibex at a distance it is relatively easy to determine foraged trees or shrubs, but not grasses or herbs. These smaller plants can be detected when visiting the place of foraging after the observation. **b** shows such a place with foraged *Alchemilla rothii* (some remaining leaves are also well visible) and *Arabis alpina* (with white flowers);**d** a common species within the *Festuca-macrophylla*-long-grass vegetation is *Simenia acaulis*, a plant heavily foraged by Walias and Klipspringer, as visible from Table 38; a place where this plant species has been eaten is shown in **f**

remarkable. The large soft, almost spongy leaves of *Lobelia rhynchopetalum* are a convenient and abundant food, although it is certainly of low energetic value and the leaves are apparently poisonous due to alkaloids (presumably Lobelin and ? Gessner 1953; Karrer 1959). It seems probable that the preference of the Walias for old yellow leaves has to be seen in connection with these alkaloids. Possibly the poison is more concentrated or exclusively present in green leaves. The following also supports this supposition. In Simen I was told by the farmers that one gets intoxicated if green Lobelia leaves are added to the fire and the head is kept in the smoke. From the spiny, attractive and everlasting small shrub *Helichrysum citrispinum*, the ibexes nip off the tips of the twigs and shoots with flowers, thus indicating that it is an apparently nutritious food. As a last detail to Table 37 I should like to mention that I only twice saw a Walia ibex drink water. Both times it was a very short action.

Table 38 refers to the feeding observations made at the place where the animal was actually foraging, as described on p. 65 (Chap. 7.5). The table provides information about the readiness to feed on the listed plant species and considers separately (a) Walia males only, (b) mixed groups of Walias including groups of young only, (c) Walia females and kids including young and (d) Klipspringers. For each class or species data are given separately for feeding place in the ericaceous belt and in the afroalpine mountain steppe (see p. 15, Chap. 3). In eight columns (the two vegetation belts considered in the above four classes a–d) the denominator shows how often a plant was found in sufficient abundance within the 50-cm range of foraging (p. 65), and the indicator records how often the plant was actually taken by the animals. For example, the figures 4/4, 3/6, 0/5 indicate respectively that the plant was always foraged, foraged in half the cases, or never foraged. Comparing the Walia ibex and the Klipspringer I should like to mention the following. Trees and larger shrubs are almost absent in the columns for the Klipspringer. *Erica arborea* has a high score for the Walias, but in individual feeding observations not shown in the table, the actual bite numbers remained low. In most cases the Tree heather did not average 1/10 of the counted bites per place. *Helichrysum citrispinum*, presumably an important food for the Walia ibex according to Table 37 and the table described here, was not seen to be taken by the Klipspringer. Grasses were foraged by both species quite often. For Gramineae, Cyperaceae and Juncaceae a summarized value was also calculated. With reference to the bite numbers counted at each place, but not shown in the table, in the Walia ibex classes there was no apparent difference between the animals observed in the ericaceous belt and those in the mountain steppe. In contrast, in the Klipspringer the average grasses foraged per observation was 0.14 in the ericaceous belt, but 0.41 in the mountain steppe ($t = 2.56$, df.: 24, $P < 0.02$). In the table this tendency is indicated by the higher score in the steppe (13/13 as compared with 7/13). The other monocotyledonous plants such as *Merendera*, *Romulea* and *Cyanotis* were mainly foraged by the Klipspringer, but with moderate frequency. The rather abundant *Alchemillas* were eaten by both species. For the larger species *Alchemilla rothii* I was originally expecting a high score for the ibex, as this plant was often declared by local people to be the favoured Walia food. In fact it was often ignored, but on a few occasions between September and March the plant was heavily foraged. At the observation where I was able to count the highest number of bites, there were 571 affecting this

Table 38. The readiness of the plant species to be foraged by Walia males, mixed groups of the Walia ibex, Walia females and by the Klipspringer. The data are based on examinations at the feeding place (p. 65, Chap. 7.5). Observations were carried out separately for the lower ericaceous belt and the upper mountain steppe. The denominator indicates in how many observations the plant was found within the 50-cm range of foraging. The indicator stands for the number of observations when the plant was actually seen to be taken by the animals. Plants within the 50-cm range which were not selected were considered in the denominator only, if they were relatively abundant and covering about 1% – 5% of the place (see also Fig. 40)

Type of vegetation	Animal class							
	Walia males		Walia mixed groups		Walia females		Klip-springer	
	Observation places Ericaceous belt (Erica) and mountain steppe (Steppe)							
	Erica	Steppe	Erica	Steppe	Erica	Steppe	Erica	Steppe
Trees and larger shrubs:								
Erica arborea	4/5	1/1	2/2	4/4	1/2	2/2		
Hypericum revolutum			1/1	1/1	0/1	0/1		
Hypericum sp.	1/1							
Rosa abyssinica	1/1							
Lobelia rhynchopetalum	4/4	6/6		5/6		1/1		3/4
Otostegia rependa	1/1							
Small shrubs:								
Blaeria spicata	0/2	0/1						0/1
Bartsia sp. (yellow flowers)				1/4		0/1		
Bartsia petitiana						0/1	0/1	2/2
Bartsia kilimandscharica		0/1		0/1		1/1	0/1	
Clematis simensis								
Clematis sp.	1/1							
Hebenstretia dentata	0/3		0/1	1/1	2/3	0/1	1/2	
Helichrysum citrispinum		2/3		9/9		3/3		0/3
Helichrysum cymosum	0/2			0/2	0/2			1/1
Helichrysum foetidum		1/1		0/1				0/1
Helichrysum odoratissimum	0/1							
Helichrysum splendidum				1/1		0/1		
Helichrysum gerberaefolium	0/1							
Helichrysum horridum				1/1			0/1	
Helichrysum formosissimum		0/1						
Rumex nervosus	0/1				0/1			
Pteridophyta:								
Asplenium adamsii		0/1	2/3	0/1		1/1	1/2	
Asplenium abyssinicum						0/1		
Cheilanthes farinosa			1/2				0/1	
Adianthum thalictroides			0/1					
Various Lichens:	4/4		2/2		2/2			
Gramineae/Cyperaceae/ Juncaceae:	5/8	10/14	3/4	9/14	1/4	7/7	7/13	13/13
Festuca macrophylla	3/6	2/12	0/2	3/12	0/2	6/6	0/4	4/12
Festuca abyssinica		1/1	1/1			1/1	1/4	1/2
Koeleria convoluta		1/2	1/1	0/2		0/1	1/1	0/1

Table 38 (continued)

Type of vegetation	Animal class							
	Walia males		Walia mixed groups		Walia females		Klip-springer	
	Observation places Ericaceous belt (Erica) and mountain steppe (Steppe)							
	Erica	Steppe	Erica	Steppe	Erica	Steppe	Erica	Steppe
Danthonia subulata	1/3	1/1	0/1	1/3	0/1	0/3	0/1	8/9
Poa simensis	1/2	4/5	2/2	2/3		1/1	3/6	6/8
Poa leptoclada		3/3						1/2
Poa schimperiana	0/1			1/1	0/1			1/1
Pentaschistis pictiglumis	0/1			0/1				2/3
Andropogon amethystinus	0/4		0/1	0/1	1/3		0/4	1/4
Andropogon abyssinicus	1/1		1/1				3/3	
Agrostis sp.	1/1					1/2		
Exotheca abyssinica							0/2	
Aira caryophyllea	1/2	1/1		0/1	0/1		0/6	3/4
Vulpia bromoides		1/1		1/1			2/8	1/1
Bromus sp.						0/1	0/2	0/1
Carex monostachya		2/10		3/5		1/3		2/5
Luzula abyssinica	0/1	1/2		1/3		0/1		5/6
Juncus capitatus							0/2	
Iridaceae/Liliaceae:								
Hesperantha petitiana		0/1						
Kniphofia foliosa	0/1	1/1			0/1			
Merendera abyssinica		1/1	0/1				4/5	0/1
Romulea fischeri	0/1	1/1	0/1	0/1			4/8	0/1
Orchidaceae:								
Disa sp.							1/1	0/2
Commelinaceae:								
Cyanotis barbata	0/1	0/1	1/2			1/1	6/12	1/4
Commelina africana							0/3	
Ranunculaceae:								
Ranunculus oreophytus		0/1				1/1		
Crassulaceae:								
Crassula sp. (*alba*)		0/1					3/3	
Crassula sp.							1/1	
Crassula pentandra		0/1	0/1	0/1		0/1	0/5	0/3
Sempervivum sp.		0/2		0/2		0/2		
Umbilicus botryoides			1/1			0/2	1/1	
Saxifragaceae:								
Saxifraga hederifolia		0/3		1/4		1/3		
Rosaceae:								
Alchemilla (*abyssinica*) *rothii*	1/2	1/5	1/2	5/9	0/1	3/6	0/1	1/3
Alchemilla sp. (small)	0/2	1/10		2/5	0/1	1/4	2/7	2/9
Papilionaceae:								
Trifolium sp.								0/1
Trifolium sp.						0/1	0/1	0/1

Table 38 (continued)

Type of vegetation	Walia males		Walia mixed groups		Walia females		Klip-springer	
	Erica	Steppe	Erica	Steppe	Erica	Steppe	Erica	Steppe
Trifolium cryptopodium							0/3	0/1
Trifolium acaule							1/2	
Trifolium arvense					0/1		0/1	
Cruciferae:								
Arabis alpina	0/1	6/9	1/1	6/10		3/5	0/1	4/4
Arabidopsis thaliana		0/2	0/1				0/1	
Cardamine hirsuta		0/2				1/2	0/1	
Geraniaceae:								
Geranium ocellatum			1/1		0/1			0/1
Erodium moschatum							0/1	
Polygalaceae:								
Polygala erioptera	1/1				1/2		0/2	
Polygala sphenoptera							1/1	
Umbelliferae:								
Anthriscus silvestris		1/1						
Ferula communis	0/1		1/1		1/1		1/2	
Haplosciadium abyssinicum							2/3	
Euphorbiaceae:						0/1		
Caryophyllaceae:								
Cerastium octandrum	0/1		0/2	0/3	0/2	0/5	6/10	4/6
Dianthus longiglumis				1/1				
Sagina abyssinica		0/1		2/4		2/2		4/6
Silene sp.				0/2	0/1	0/1		
Silene flammulifolia	1/1	0/1		1/1		1/2	0/1	0/2
Minuartia filifolia	0/2			0/2	0/1		0/1	
Uebelinia abyssinica							1/5	
Primulaceae:								
Primula verticillata		0/1						
Gentianaceae:								
Swertia erythraeae		0/1		2/4	0/2	0/1	3/3	1/1
Swertia sp.		1/4	1/1			1/2	2/3	3/6
Boraginaceae:								
Myosotis vestergrenii			0/1	1/2		0/1		
Myosotis (abyssinicus)				1/3			1/5	0/1
Cynoglossum								
Labiatae:								
Thymus serrulatus	0/1	0/2	0/1	0/3	0/3	0/1	0/10	1/2
Satureja pseudosimensis		0/2	0/1	0/6	0/2	0/1		1/5
Satureja punctata	0/3	0/1	0/4	0/2	0/3		2/7	
Nepeta sp.	0/1							

Table 38 (continued)

Type of vegetation	Animal class							
	Walia males		Walia mixed groups		Walia females		Klip-springer	
	Observation places Ericaceous belt (Erica) and mountain steppe (Steppe)							
	Erica	Steppe	Erica	Steppe	Erica	Steppe	Erica	Steppe
Salvia merjamie								0/1
Plectranthus		0/1					0/1	
Scrophulariaceae:								
Celsia sp.				0/1				
Veronica glandulosa							0/1	
Plantaginaceae:								
Plantago afra L. var. *stricta*	0/1	0/1		1/1	0/1		7/11	
Rubiaceae:								
Galium aparine			0/1				2/6	
Galium spurium			0/1				0/2	
Valerianaceae:								
Valerianella microcarpa		0/1	0/1	0/1	0/1		3/10	1/2
Dipsacaceae:								
Scabiosa columbaria			2/3		2/3		2/2	
Simenia acaulis (id. with *Dipsacus pinnatifidus* in Iwamoto, 1979, p. 285)	3/3	9/9		10/10	1/1	3/5		6/8
Compositae:								
Anthemis tigreensis								
Bidens sp.	1/1		0/1				1/3	
Coreopsis boraniana	0/1		0/1		1/1			
Cotula abyssinica			0/1				0/3	
Crepis newii			1/1	0/1			5/6	3/3
Crepis oliveriana							1/1	1/1
Conyza tigreensis							1/1	
Dichrocephala chrysanthemifolia		0/1				0/1	1/2	
Filago spathulata							0/4	
Haplocarpha rueppellii				1/1		0/1		
Launaea cornuta			0/2			0/2	0/3	
Senecio sp.								0/1
Siegesbeckia abyssinica	0/1							
Tolpis altissima	1/1		1/2		1/2		3/3	
Phagnalon nitidum	0/1				0/2			

Alchemilla out of a total of 629 bites. The Papilionaceae seem to be avoided. Among the more frequent and therefore quantitatively important plants, the following were foraged by both ungulate species: the long-flowering cosmopolitan *Arabis alpina*, *Swertia* sp., *Scabiosa columbaria*, *Simenia acaulis*, which, being included 20 times for the Walia ibex and 6 times for the Klipspringer, was the plant species that provided the highest number of bites per observation, as well as *Tolpis altissima*. The following plants seem to be foraged mainly by the Klipspringer: *Crassula* sp., *Cerastium octandrum*, *Swertia erythraeae*, *Plantago afra*, *Valerianella microcarpa* and *Crepis newii*. It seems noteworthy that none of the listed Labiatae were ever seen to be foraged by Walias (with the exception of the shrub *Otostegia rependa*), and only rarely by the Klipspringer. To summarize, in the comparison between the Walia ibex and the Klipspringer there is a wide overlap in the species selection, with however some remarkable distinctions, the ibex with a preference for trees and shrubs including spiny species such as *Helichrysum citrispinum* and *H. horridum*, the Klipspringer with a list of mainly small and soft herbs. Qvortrup and Blankenship (1974) report the Klipspringer living in a bush-covered area surrounded by dry pasture-land as selecting also very few grasses only, but – besides herbs – also several woody plants.

For the comparison between different Walia ibex classes in Table 38 I do not see any clear evidence of a distinguishing behaviour in the selection of plant species, with the possible exception of *Lobelia rhynchopetalum*, which is more represented in the male column (see also Table 37).

Dunbar (1977a, 1978a) and Iwamoto (1979) have investigated the feeding habits of the Gelada baboon and from their results this primate may be compared with the Walia ibex and Klipspringer discussed here. Both authors clearly consider the Gelada as a grass-eater. Dunbar (1978a) averages the value as 92.7% of grasses in the diet, and reports for comparison 10.8% for the Walia ibex and 17.3% for the Klipspringer. But there were apparent dissimilarities between different individual Geladas and between the seasons (Iwamoto 1979). These variations also include a preference for different grass species. Bulbs of *Merendera* and *Romulea* were dug throughout the year, but *Trifolium* roots, stems and leaves were taken mainly in the dry season, as well as leaves and flowers of *Helichrysum citrispinum*. *Swertia*, *Satureja* and *Dipsacus* plants (*Simenia acaulis*) were not eaten. In comparison with the two ungulates, the inverse behaviour with regard to *Trifolium* and *Simenia* is remarkable (Table 38). The data quoted above are consistent with my own more meagre observations on the feeding habits of the Geladas.

It was next examined whether the diversity of the food varies between different classes and species. In the summarized Table 39, the average number of different plant species per observation, and the average amount of the mainly foraged plant species are given. This second figure may be considered a reciprocal measure for the evenness. The higher the relative number of bites of the leading plant species, the smaller the evenness. As Table 39 shows, in both measures there is no significant difference between the three Walia classes; however, the comparison of the pooled Walia classes with the Klipspringer shows apparent distinctions. The Klipspringer forages more different plant species per observation and there is usually no clearly dominating plant species as the in Walia ibex. This is consistent with the observed behaviour and the strong field impression that the Klipspringer collects its food tipwise, clipping tiny portions only here and there. With such a feeding tactic it is possible to select highly digestible forage such as flowers, fruits and sprouting shoots with great precision. However, the concentration on such tiny fractions of the plant biomass that is actually available can be suitable only for small-bodied animals like Klipspringer (Geist 1974b). With regard to the Walia ibex and

Table 39. Diversity of food of Walia ibex males, mixed groups of Walia ibex, Walia ibex females and the Klipspringer according to careful observations at the foraging place. The average number of different plant species selected per observation ($\bar{x}$), the number of considered observations (n) and the standard deviation (s) are given. The observed numbers of plant species was compared with an F-test, first between the three Walia ibex classes only and second between Walias and Klipspringers. Additionally, and treated in an analoguous manner, the percentage values of the number of bites of the mainly foraged plant per observation were compared between the classes and species. Thus the first diversity measure considers simply the number of species, the second estimates the evenness, but in a reciprocal sense. In both measures, only observations with at least ten bites were considered

		Walia males	Walia mixed groups	Walia females	Klip-springer
Number of	$\bar{x}$	4.1	5.3	4.8	6.5
different	n	16	18	10	24
plant species	s	2.0	2.9	2.9	3.0
Comparison between					
Walia ibex classes		df 2; 41	F=1.01 n.s.		
Walia and Klipspringer		df 1; 66	F=6.00, P<0.05		
Amount of the mainly	$\bar{x}$	68.8	69.9	74.8	45.8
foraged plant species	n	16	18	10	24
in %	s	20.1	20.8	14.1	16.8
Comparison between					
Walia ibex classes		df 2; 41	F=0.32 n.s.		
Walia and Klipspringer		df 1.66	F=28.82, P≪0.01		

prediction 9, the information given with Tables 38 and 39 does not reveal differences between males and females. There is no significance in the pattern that the amount of the mainly foraged plant species in percentages is slightly higher for females, although they have taken – on the average – few more different species than males.

In considerations on the feeding tactics, the flowering period was estimated to be of importance. For the same animal classes as above, Table 40 deals with the foraging intensity, separated for different growth stages of the selected plants. The tendency that results from the codified bite numbers given in the table coincides surprisingly well with the prediction that Walia males favour the freshly growing plants, whereas females look for fruits (χ^2:120.2). However, the validity of this result with regard to the prediction remains doubtful as it was written without quantitative information, e.g., on the energy that males and females actually require over the seasons and the nutrition content of the foraged plants in different growth stages (see p. 145). In any case the difference between males and females in Table 40 reveals a tendency to select different phases of plants.

For the Klipspringer the pattern seems more evenly distributed than that of the Walia females, even though a small preference for the fruit phase may be supposed. However, and interpreted to reveal relative differences in the behaviour of the two ungulates, the Klipspringer tactic is more likely to select specific parts of plants within a wide number of different species than to look for a specific phase such as sprouting or fruiting within a restricted diet. According to my data, this behaviour

Table 40. Comparison of the growth stage of the foraged plants with regard to the flowering period that are selected by the Walia ibex males, by mixed groups of the Walia ibex, by the Walia ibex females and by the Klipspringers. For these classes and for six different stages the codified number of counted bites per plant and per examination at a feeding place was given. The same code that is based on the log 2 scale was applied as introduced in Chapter 8.1 (p. 68). For each foraged plant the phase in comparison to the flowering period was determined (see Table 2). The three flowering phases were determined as follows: Plants with an observed flowering period of 1 or 2 months only, all data within this short period were declared as central period. A flowering period of 3 months was divided into begin, central period and end, and in all plants with a longer flowering period, the first 2 months were considered the begin, the last two months the end and the remaining months in between the central period. The distribution pattern for the four animal classes differs with high significance, as was shown with a contingency-table test ($\chi^2 = 120.2$, df. = 15, P ≪ 0.01)

Growth stage of the plants being foraged		Walia males	Walia mixed groups	Walia females	Klip-springer
1–3 months before flowering		44	53	14	49
In flowers	Begin	51	28	6	40
	Central period	36	37	25	55
	End	27	40	28	63
1–3 months after flowering		25	37	64	64
Other seasons		15	57	13	29
Total		198	252	150	300

is the case for the Walia ibex. The fact that the Alpine ibex in summer shows a preference for meadows where the snow has melted recently indicates possibly that the growth stage of the plants being foraged is also important in this other subspecies of Caprinae (Nievergelt 1966a).

A higher per-gram metabolic rate may also be achieved by utilizing the foraged plants more carefully. While ruminating, the food is given back to the mouth portion-wise, chewed a certain number of times and then swallowed again. I have now considered the number of chewings per portion as a measure of the care and effort to utilize the food. Distinguished according to three Walia ibex classes: i.e., females, various young and males of at least 3 years, 21 average chewing values were compared. For females, the mean was 78.0, for the young 82.0 and for adult males 62.25. The difference is significant among the three Walia classes (df 2/18, F = 4.25, P <0.05); in the comparison females and the young pooled opposed to adult males, a higher level of significance was achieved (df 1/9, F = 8.37, P <0.01). The result indicates that females (and young) with their assumed higher per-gram metabolic rate may in fact be more careful in utilizing their food. It must be added in this context that a well-developed ability in disclosing plant material is to be expected for the ibex, as goats and sheep seem to be generally adapted to rather rough food quality. For the Klipspringer I have only two average values: 44 and 48, thus apparently a lower number, and this possibly indicates an adaptation of this small antelope to a relatively softer food.

Reviewing the results given in this chapter on food selection, the data on feeding observations at a distance (Table 37), those on the phase in which the plants are selected and the above on ruminating frequencies seem to be consistent with the prediction that females of the Walia ibex, in comparison with adult males, have a behaviour to satisfy relatively higher protein and energy requirements.

16.7 Conclusions and Comparisons with the Situation in the Alpine Ibex

In the earlier sections of this chapter it was shown that the data available are generally consistent with the inferred behaviour of the different Walia ibex classes. However, not all the data are conclusive. Predictions 1 to 3 on peculiarities in the ecological behaviour of males and females, prediction 6 concerning the question of dominance, and prediction 7 about the marching order were confirmed with a high degree of reliability. Prediction 9 on the feeding strategies is supported with a fair amount of evidence, but from the scant data on predictions 4 and 5 concerning home-range size and the visitors' role of adult males, only indications of low validity could be given. As to prediction 8, expecting efforts of the kids to prevent further copulations of their mothers, there was no evidence at all (see Chap. 15.3, p. 144). In this situation we must assume that the true social system that was attempted in an approximate sense and a wide range of confidence, has the shape described in Chapter 15. No or no clear answer is possible yet, for instance to the following three questions:(1) What degree of flexibility of the system exists? Keeping in mind that the dominating position of adult males may only be achieved after several years of experience a high flexibility seems likely. (2) Do we have to assume that there will be one male only in the dominant role within a particular area, or, e.g., in the case of increased population density, does the system permit largely overlapping ranges of various "dominant" males? What are further possible density-dependent effects? (3) Do females in oestrus actually remain within their ordinary home range system and activity pattern or do they search actively for a partner to mate? In the social system described above, all responsibility for the timely meeting of males and females is given to the male. However, a limited contribution of the female in this activity does not contradict the findings and is not to be excluded a priori. The difference in the average habitat between females in female-juvenile groups and females in mixed groups as shown in Fig. 38, even though it is of a minor degree, could be explained as being partly caused by a possible active change in the habitat selection of females that are willing to mate. In this context, I should like to recall also the inference that females have possibly smaller home ranges than males. A small size of the females home range obviously makes the search more easy for the male. Evidently a study with individually known animals would be required to narrow the field of speculation.

The social system as deduced in Chapter 15 is linked with the environmental situation in the afroalpine climate in Simen. It cannot be expected, therefore, that other ibexes are organized in the same way. In order to carry out a fruitful comparison with other Caprinae, that could lead to a broader basis of understanding, it is essential to consider basic differences with the temperate latitudes that house most other ibexes. From this starting point, a comparison with the Alpine ibex is offered below.

The most fundamental difference in the habitat of the Alpine ibex is the existence of clear-cut seasons with a winter that causes fundamental limitations to animal life. In winter, large zones of the range are uninhabitable, and a population is often separated, even isolated, into several distinct suitable winter ranges and concentrated therein depending on actual snow conditions and population density. Major changes in the conditions of the habitat, e.g., the vegetation, take place between spring and autumn. We have thus distinctive ranges for each of the seasons. The winter, just described as being a limiting influence in the distribution of the animals, dictates also that spring be the season of birth; thus the annual reproductive cycle is scheduled into a seasonally distinct pattern (Nievergelt 1974).

Out of these facts a number of conclusions can be drawn: In the Alpine ibex there is a long period in the year's cycle with no females in oestrus; thus males are not forced to control and visit female groups in these periods; they may therefore concentrate on quarrels and other activities for establishing rank among competing males. Determined by the parturition season and the duration of pregnancy, rutting takes place in winter. Thus females are simultaneously in oestrus when migration possibilities are limited, and the animals concentrated in suitable places (Nievergelt 1966a; Daenzer 1979). In this obviously different situation the tactic of old males cannot be the same as in Simen. Circulating within a large home range is not possible during the rut due to snow conditions and it is also unsuitable because between the associations of animals in the various winter ranges the rut and the oestrus cycles are largely synchronized. For dominating males a more appropriate tactic seems to be to reside within a large and qualified winter range, and therein to control the various females that are associated there in often reasonable numbers. Findings in the Alpine ibex seem to support this prediction (Nievergelt 1966a, 1967).

There is nevertheless evidence of similarities in the distribution pattern of the two ibexes. As predicted, and as also probable according to the data, adult Walia males have larger home ranges over the year than females. The same seems to be true for males in at least a reasonable number of colonies of the Alpine ibex in Switzerland such as at the Augstmatthorn, the Justistal, the Piz Albris, in the Val Bever and the Safiental (Nievergelt 1966a). However, the reason for this phenomenon is more likely a further similarity in the ecological behaviour than the visiting and controlling of various female groups. Males in the Alpine ibex in contrast to the females – like the Walia ibex – do not frequent extreme habitats (such as steep slopes) as exclusively as females do. According to my own observations, there is good evidence that these two statements on different range size and type of habitat are also true for the Wild goat in Iran. With regard to the Walia ibex we are disposed to speculate: The higher tolerance of males in utilizing different habitats, and this could be a behaviour already adapted before the invasion of afroalpine regions, has promoted the tactic of adult males to range in a larger area and to visit female groups therein.

It seems in fact probable that all such apparently fundamental differences in the social system of the Walia ibex and the Alpine ibex, e.g., particular ranges for males and females, as well as for different seasons, reproductive cycle etc., may have been caused exclusively by the difference in climate and habitat. In considering the various environmental constraints, a high degree of relationship in the social system of the two ibexes is immediately apparent.

Summary to Chapters 15 and 16

This part of the study deals with the social and ecological behaviour of different sex-age classes of the Walia ibex. Other species living in Simen are only considered in Chapter 16.6 on food selection.

In an approach to understanding the social system of the Walia ibex, the tactics of the class members are discussed. In selecting places and habitat, males, as compared to females, seem to be ordered mainly according to social conditions and are apparently less governed by habitat factors and vice versa. In the habitat map (Fig. 35, p. 126) old males are most often observed on hectare fields of low habitat value but rarely on fields of high value. The behaviour of females and kids is reciprocal.

According to a reproductive cycle with rutting throughout the year – although with a definite peak – males join female groups during all seasons: young males more steadily, old males more loosely. Old, dominating males are inferred to monitor and check various female groups. A marching order with females in front and males behind seems to indicate a temporary active subordination of the males in these groups and to support such a monitoring tactic. There is some weak evidence for old males occupying larger home ranges than females and young males. Such a behaviour would also correspond to the old males' role of a circulating visitor.

Data on food selection indicate obvious distinctions between the Walia ibex, the Klipspringer and the Gelada baboon, but more subtle differences between males and females of the ibex. There seems to be a tendency for the females to forage relatively more energy and protein per body weight than males. According to my observations males more often feed on the leaves of *Lobelia rhynchopetalum* and more rarely of *Helichrysum citrispinum*; they apparently favour freshly growing plants whereas females seem to prefer the fruit phase. When ruminating, males chew less often per portion than females.

In a comparison to the Alpine ibex it is concluded that the differences in the social system between the two species are explicable as effects of the differences in climate and habitat.

Conservational Outlook

17 The Simen, an Ecosystem in Danger

Various reports and papers have been written concerning the uncertain future of both the natural flora and fauna of the Simen mountains as well as of the human population, including their culture and economic situation. There is an apparent variety of problems, but in fact, they are all firmly linked, and some are direct consequences of others. In the first part of this chapter I shall briefly review those problems that have arisen and require consideration on a human time scale. They have already received some attention. In Ethiopia, in the last ten years the Simen mountains have been recognized as a national asset, and many efforts have been made by the Government and the Ethiopian Wildlife Conservation Organization in cooperation with international organizations such as WWF and IUCN to preserve this area. This is still far from being sufficient to achieve the goal, but one must remember that the country is vast and the Simen mountains remote. This remoteness, however, has not discouraged the scientific interest, as is documented by an already remarkable literature on this area (see Schaerer 1979) and by UNESCO in listing the Simen Mountains National Park as World Heritage Site (World Heritage Committee Meeting in Washington, D.C., September 1978). In the second part of the chapter an attempt at taking a long-term view is made in that some expectations are discussed on an ecological time scale. Bearing in mind the various short- as well as long-term problems, in the last section of the chapter, suggestions are made with the aim of helping maintain or increase the value of this unique ecosystem for the benefit of all wildlife and the indigenous people depending on it.

Among the immediate problems I wish to summarize firstly the several threats to the whole natural community. With regard to the endangered Walia ibex they are of an indirect character, but very clearly they must receive principal attention. Direct threats to the ibex will be considered subsequently.

As was described in the section on human utilization of the area in Chapter 3 (p. 24), the threats to the natural community are mainly caused by the increase in the human population, the scattered character of the settlements, that cut the natural habitats into small fractions and by the burning of forests and the loss of soil due to ploughing of slopes without forming terraces and subsequent erosion. In order to estimate which habitats and species are particularly endangered, the differentiation in land use by man must be considered. With reference to Table 11 (p. 71) this question was discussed on p. 26. Undoubtedly, the most apparent deficiency has occurred within the original montane forest. This rich habitat has disappeared even within various parts of the park. Today, valuable trees such as *Juniperus procera* have become rare in all of the Simen mountains. Among the larger mammals that occupy this belt, we must mention mainly the Bushbuck and the Colobus monkey (see Fig. 20, p. 45 and Table 25, p. 112); the occurrence of both species in Simen is linked with this type of forest, but the steeper slopes are also part of the Walia habitat.

Among the direct threats to the Walia ibex we must discuss hunting or poaching and also cross-breeding with domestic goats. Both problems are easier to manage successfully than the heavy burden of indirect threats due to habitat loss mentioned above. However, these problems are serious too and should not be neglected. We must remember firstly that the inhabitants of Simen, as other indigenous people in Africa, have always regarded wildlife as a source of animal products (Talbot 1966, see also p. 27 and Fig. 13). In all conservation measures taken one must be aware that the sudden prohibition of a traditional resource is more difficult to understand than it would be for hunters in Europe or North America where wildlife is commonly regarded as a source of sport hunting only. Nevertheless, it is the impression of all the recent Park Wardens (J. P. Müller, P. Stähli, H. Hurni) that poaching of the Walia ibex did not occur often within the park borders, possibly even less so than during our field work in 1968/69 (see p. 27). In fact, the protection of the Walia ibex seems to be the most readily accepted regulation of all by the people. It is very clear, however, that any serious interruption of the control may well lead to the extermination of the Walia ibex in the Simen (see also Nievergelt 1972b). Due to its small size, the natural habitat is vulnerable. As a result of the Italian invasion, a remarkable number of rifles has remained in this area. This fact has increased the efficiency of the hunters. Probably, this change in the weapons and the shrinking of the natural habitat are both reasons why the population of the Walia ibex has dropped to its present alarming low number.

The aforementioned problem of cross-breeding with domestic goats may seem far-fetched to most readers. However, we must realize that, according to our knowledge, any hybrids produced by cross-bred species or subspecies within the genus Capra are fertile (Couturier 1962; Gray 1972; Heptner et al. 1966; Herre 1958). Moreover, we may assume that the possibility of hybridisations between Walia ibexes and domestic goats increases when the number of lone and scattered individuals is high, in either one or both species. Thus, a low population density in the Walia ibex is supposed to favour such a genetic misfortune (see also Nievergelt 1966 and therein p. 76 and Fig. 40). In order to preserve the genetic purity of the Walia ibex one of the protective measures must be to keep the domestic goat out of the Walia ibex range.

Despite the burden of the problems that require immediate attention and despite our limited opportunities to estimate future developments, we must endeavour to forecast the result of the various measures on a long-term scale. For 19 nature reserves in East Africa Soulé et al. (1979) have calculated four different models, in order to predict how many of the large mammal species will be lost in 50, 500 and 5,000 years. With all the models it is assumed that the reserves will be maintained as reserves, but isolated from each other in the future, and that the dynamics of the faunal collapse will be similar to the dynamics of extinction of large mammals on seven land-bridge islands in Southeast Asia in the Post-Pleistocene period. These assumptions would seem to be realistic, as – for the time being – several reserves actually show a clear tendency to be surrounded by cultivated land and are thus isolated. The predictions are based on the island theories, but for the special case, where, due to the isolation, the possibility of new invasions is zero (Brown 1971). According to the two intermediate models, considered to be the more realistic ones, a reserve of the approximate size of the Simen Mountains National Park will lose between 5%, 35% and 85%, and 25%, 65% and 90% respectively of their large mammal species in 50, 500 and 5,000 years after being isolated. This pessimistic outlook is calculated assuming that any human interference is absent. Of course, it is unknown to what extent all assumptions that have led to the models of Soulé et al. (1979) are justified. However, the island theories have been successfully applied to various terrestrial ecosystems. Today, the biological consequences of man-made insularity in a natural terrestrial community caused by the creation of inhospitable terrain in the surrounding area must be considered a fateful and non-compensatory burden for the respective community (Diamond 1975a; Mac Arthur and Wilson 1967; Terborgh 1975). From such studies we also know that the different animal species will not become extinct at random (Diamond 1975b). Generally, those animals are most endangered that have large home ranges, are restricted to very particular habitats, and occur in low numbers also in the untouched natural situation; e.g., most predators. Seen from here, the Walia ibex is not likely to be the first to become extinct in the situation of absolute man-caused insularity. Other mammals, like Serval cat, Simen fox and Colobus monkey, would possibly disappear first. However, it would be a rather optimistic view to believe that, in a time scale such as 500 years, the Walia ibex would survive in a range strictly limited to the actual Park borders.

Having in mind the various immediate threats to the unique afroalpine community, to the Walia ibex and to the economic basis and culture of the people, and considering also the long-term aspect we now have to ask: what sort of measures are suitable to help preserve the Simen?

Very clearly, the Park cannot be seen in isolation. In contrast, a comprehensive land-use planning is required in the whole of Simen that includes the elaboration of management alternatives for the different areas of interest to conservation. It is the leading idea of the Pro Semien Foundation, to study carefully the aspects of nature and culture in a wide angle view, and, stemming from this, initiate technical projects on a modest scale suitable to ameliorate the situation without causing any damage to natural and cultural values. The Pro Semien Foundation is an ad hoc organization in Switzerland formed by scientists of different disciplines, interested institutions, towns and private persons in order to study, propose and support activities aimed to preserve nature and man in Simen. So far, the Ethiopian Government has been carrying much of the conservation activities in the Simen Mountains National Park. On the other hand, it is obvious that Ethiopia cannot finance the required comprehensive project within a reasonable space of time. Foreign aid and support of international organizations are indispensable.

A number of regulations that refer to the National Park area are purely conservationist in character, almost self-explanatory and totally or partly realized already: the protection of wild animals and plants, regular controls in order to prevent further habitat destruction and poaching, people and domestic stock kept outside the park border, although within the park possibly on certain permitted paths. Movements within the park are to be directed and controlled. However, as indicated already with the attempt at a long-term forecast, some conservation regulations are essential also outside the park boundaries. A change of the entire surrounding area into agricultural land would drastically reduce the survival chances of the natural community as such and this tendency would inevitably serve to accelerate the rate of extinction. Corridors of natural habitat ought to be maintained in order to preserve a free interchange of populations within the various and largely scattered natural or seminatural habitats within the whole mountain range (Terborgh 1975; Wilson and Willis 1975). An extensive level of agricultural use that does not exclude all natural plant species can give some additional support. Thus, a heterogeneous system of settlements, of agricultural areas at different levels of cultivation intensity and of natural communities that have connecting arms or fingers appear as an optimal land-use system for the mountain area in Simen outside the National Park. Within such a system, the National Park itself represents a core area, where all vegetation belts and types of habitat will be preserved in a natural state and linked in their original pattern. Outside the Park border corridors of particular habitats, e. g., Tree heather woodland, will be the major elements of interest to conservation. As at the present time most of the "corridors of the natural habitat" have already disappeared – in wide areas even in the "puna"-region – an approach for the re-establishment of such zones should be undertaken and included among other programmes. Fortunately, the topographical situation and the difference in the influence of man to the various habitat types seems to favour such an approach. Climate, steep slopes, rocks and gorges inhibit naturally or at least retard the advancing agriculture in some areas although the cutting and/or burning of forests, the first step on the way to denudation and devastation of natural habitats, is nowhere sufficiently prevented by the sole topographic situation.

The aim to preserve or even enlarge natural habitats within the Simen Mountains National Park and also in some further remnants outside the park boundaries is extremely ambitious, but it must have its place within a justifiable land-use policy, and it cannot be considered unrealistic. As the whole region of the Simen mountains is wide but not endless, it is still capable of being monitored. It is clear, however, that the various remnants of natural habitat cannot be preserved with conservation measures alone. One of the most immediate needs to be achieved, apart from the obligation to lower the pressure of the human population, is that methods of land use around the Park area need to be modified so that erosion can be stopped, and the human population is no longer depending continuously on new land resources (Nievergelt 1973; Winiger 1976). It is urgent that a reafforestation programme should be accelerated and completed. Wood is essentially required as a building material, for cooking, and, due to the high altitude, also for heating. If the remnants of the natural forests are to be saved, a renewable resource must be available. The quickly growing *Eucalyptus globulus* or *Pinus* sp that are planted in most Ethiopian regions to satisfy the daily demands for wood may serve as short-term solutions. In the long run, reafforestation with original species must be promoted in those areas where erosion has not yet abolished the hope of recovery (Tewoldeberhan Gebregziabher 1974). With their rich undergrowth these species provide better protection against erosion, help to regulate the water regime and serve as habitat and corridor for a variety of indigenous plant and animal species. It would be appropriate to aspire to a future situation with a moderate degree of harvesting in these natural or seminatural areas outside the park boundaries. This may be important also for psychological and educational reasons, because the planting of indigenous species for protection only and the exclusive using of exotic trees may lead to a lack of interest into the original resources. The traditional use of the natural resources should be maintained as far as possible and this includes such details as the flowers and fruits of *Hagenia abyssinica* as vermifuge, the roots of *Rumex abyssinicus* to make tea, the transformation of a dead inflorescence of *Lobelia rhynchopetalum* into a "throwaway" raincoat or the utilization of hand-rolled tussock grasses as ropes. The inhabitants should not live with the feeling of being excluded from their natural resources.

The present situation with small and therefore vulnerable remnants of the natural habitat and a poor basis for the people living in Simen requires regular control of habitat changes, thus the establishing of monitoring programs in natural as well as in cultivated areas. The recording of such changes in whatever way are essential in order to judge to what extent the measures taken may be considered successful, or if further action must be started. The taking of regular photographs from certain selected points is a first and valuable technique. An example was given in Fig. 12 (p. 26). The regular counts of the Walia ibex from various fixed observation points as presented above are a further technique (see Chap. 9, p. 80). Counts on permanent transect lines or in test areas are a third possibility to detect slow and subtle changes in the habitats examined.

As mentioned in the introductory chapters, the Simen mountains are the only region in the world to house the Walia ibex. Success in the conservation of the last natural habitats in Simen is a prerequisite for the survival of this palaearctic ungulate in Africa. Of course, the establishment of a breeding nucleus in a zoo may help diminish the risk of a total extinction, and give some additional aid to help preserve the genetic substance of the Walia ibex. It can, however, only be regarded as an artificial means to overcome critical and relatively short periods. Thus, whether with regard to the plants, the animals or the people living in Simen, all efforts should be undertaken to preserve this unique afroalpine region. It is to be hoped that, after the national and international attention with the gazetting of a central part as the Simen Mountains National Park, and the listing as World Heritage Site, understanding and respect will also be achieved at the decisive local level.

References

Aerni K (1978) The panorama of the Imet Gogo (3926m) in Simen (Ethiopia). In: Messerli B, Aerni K (eds) Simen Mountains-Ethiopia, vol I. Cartography and its application for geographical and ecological problems. Jahrb Geogr Ges Bern, Beih 5, Geographica bernensia G8:101/102

Aeschbacher A (1978) Das Brunftverhalten des Alpensteinwildes. E Rentsch, Erlenbach-Zürich:88 pp.

Bailey AM (1932) The heights of the Simyen: natural history. J Am Mus Nat Hist 32:61-74

Batcheler CL (1968) Compensatory response of artificially controlled mammal populations. Proc N Z Ecol Soc 15:25-30

Bell RHV (1970) The use of the herb layer by grazing ungulates in the Serengeti. In: Watson A (ed) Animal populations in relation to their food resources. Blackwell, Oxford Edinburgh, pp 111-124

Bell RHV (1971) A grazing ecosystem in the Serengeti. Sci Am 225 (1):86-93

Bertram BCR (1978) Living in groups: predators and prey. In: Krebs JR, Davies NB (eds) Behavioural ecology. Blackwell Sci Publ, Oxford, pp 64-96

Blower JH (1966) Proposal for the establishment of a national park in the Simien Mountains, Ethiopia. Mimeograph, Wildl Conserv Dep, Addis Abeba: 11 pp.

Blower JH (1968a) The wildlife of Ethiopia. Oryx 9:276-283

Blower JH (1968b) Proposals for the development of the Simien Mountains National Park and other associated conservation measures. Mimeograph, Wildlife Conserv Dep, Addis Abeba: 10 pp.

Blower JH (1969) Wildlife conservation in Ethiopia. Walia 1:15-23

Blower JH (1970) The Simien-Ethiopia's new National Park. Oryx 10:314-316

Blumenthal MM (1962) Ein Streifzug durch das Semien-Gebirge in Nord-Aethiopien. Alpen 2:1-20

Bosmans R, Moreaux F (1977) Birds of the Ethiopian Begemdir Province in August and September 1973-1974. Gerfaut 67:395-412

Boswall J, Demment M (1970) The daily altitudinal movement of the white-collared pigeon *Columba albitorques* in the High Simien, Ethiopia. Bull Br Ornithol Club 90:105-107

Brown JH (1971) Mammals on mountaintops: non-equilibrium insular biogeography. Am Nat 105:467-478

Brown LH (1965a) Ethiopian episode. Country Life, London; 160 pp.

Brown LH (1965b) Africa. A natural history. Random House, New York, pp 46-61

Brown L (1966) La conservation de la flore et de la faune sauvages en Ethiopie. Nature Resour UNESCO, pp 6-10

Brown L (1969) The Walia ibex. Walia 1:9-14

Buechner HK (1960) The bighorn sheep in the United States, its past, present and future, vol IV. Wildlife Monogr. Wildl Soc, Louisville: 174 pp.

Butzer KW, Hansen CL (1968) Desert and river in Nubia; University of Wisconsin Press, Madison, Milwaukee, and London, 562 pp.

Campbell I (1974) The bioenergetics of small mammals, particularly *Apodemus sylvaticus* (L), in Wytham Woods, Oxfordshire. Thesis A E R G, Dep Zool Oxford

Carlstead K (1973) Observations on the social behaviour of a captive group of Siberian ibex (*Capra ibex sibirica*). Chicago Zoological Park, unpublished

Chatterjee S, Price B (1977) Regression analysis by example. Wiley, New York; 228 pp.

Cheatum EL, Morton GH (1946) Breeding season of white-tailed deer in New York. J Wildl Manage 10:249-263

Cloudsley-Thompson JL (1969) The zoology of tropical Africa. Weidenfeld and Nicolson, London; pp 355

Coe MJ (1967) The ecology of the Alpine Zone of Mount Kenya. W Junk Publ, The Hague; 136 pp.

Coe MJ (1969) Microclimate and animal life in the Equatorial Mountains. Zool Afr 4 (2):101-128

Coe MJ, Foster JB (1972) The mammals of the northern slopes of Mt. Kenya. J East Afr Nat Hist Soc Nat Mus 131: Nairobi; 1-18

Coetzee JA (1964) Evidence for a considerable depression of the vegetation belts during the upper pleistocene on the East African Mountains. Nature (London) 204:564-566

Coetzee JA, Van Zinderen Bakker EM (1967) Climatic changes and the stratigraphy for the upper quaternary in Africa. VIth Session Pan-African Congr Prehist Quat Stud, Dakar

Coetzee JA, Van Zinderen Bakker EM (1970) Palaeoecological problems of the quaternary of Africa. S Afr J Sci March 1970:78-84

Corbet GB, Yalden DW (1972) Recent records of mammals (other than bats) from Ethiopia. Bull Br Mus Nat Hist 22:213-252

Couturier MAJ (1962) Le bouquetin des Alpes. Edit par l'auteur, Grenoble; 1564 pp.

Crook JH (1966) Gelada baboon herd structure and movement; a comparative report. Symp Zool Soc London 18:237-258

Daenzer L (1979) Aktivitätsmuster und Zeitbudgets von Steinböcken und Gemsen im Winter. Diplomarbeit, Univ Zürich

Daniel C, Wood F (1971) Fitting equations to data. Wiley, New York; 342 pp.

Danz W (1975) Münchner Manifest zur Entwicklung von Bergregionen. In: Entwicklungsprobleme in Bergregionen. 1 Konf Club Munich, Heft 3. Alpen-Institut, München

Dasmann RF, Milton JP, Freeman PH (1973) Ecological principles for economic development. IUCN and Conserv Found. J Wiley, London; 252 pp.

Delany MJ, Happold DCD (1979) Ecology of African mammals. Trop ecol Ser. Longman, London New York; 434 pp.

Diamond JM (1975a) The island dilemma. Lessons of modern biogeographic studies for the design of natural reserves. Biol Conserv 7:129-146

Diamond JM (1975b) Assembly of species communities. In: Cody ML, Diamond JM (eds) Ecology and evolution of communities. Harvard Univ Press, Cambridge Mass, pp 342-444

Dice LR (1945) Measures of the amount of ecologic association between species. Ecology 26:297-302

Dillmann CFA (1885) Ueber die Regierung, insbesondere die Kirchenordnung des Königs Zar'a-Jacob. Philos hist Abh königl Akad Wiss Berlin, 1884

Dorst J, Dandelot P (1970) A field guide to the larger mammals of Africa. Collins, London; 287 pp.

Draper NR, Smith H (1966) Applied regression analysis. Wiley, New York; 407 pp.

Duerst U (1926) Das Horn der Cavicornia. Denkschr SNG, Fretz, Zürich, 179 pp.

Dunbar RIM (1977a) Feeding ecology of Gelada baboons: a preliminary report. In: Cluttar-Brock TH (ed) Primate ecology: studies of feeding and ranging in lemurs, monkeys and apes. Academic Press, London New York: pp 251-273

Dunbar RIM (1977b) The Gelada baboon: status and conservation. Primate conservation. Academic Press, London New York; pp 363-383

Dunbar RIM (1978a) Competition and niche separation in a high altitude herbivore community in Ethiopia. East Afr Wildl J 16:183-199

Dunbar RIM (1978b) Sexual behaviour and social relationship among Gelada baboons. Anim Behav 26:167-178

Dunbar RIM, Dunbar P (1972) The social life of the Gelada baboon. Walia 4:3-13

Dunbar RIM, Dunbar EP (1974a) Social organization and ecology of the Klipspringer (Oreotragus oreotragus) in Ethiopia. 2. Tierpsychologie 35:481-493

Dunbar RIM, Dunbar EP (1974b) Ecological relations and niche separation between sympatric terrestrial primates in Ethiopia. Folia Primatol 21:36-60

Dunbar RIM, Dunbar P (1974c) Behaviour related to birth in wild Gelada baboons (*Theropithecus gelada*). Behaviour L 1-2:185-191

Dunbar RIM, Dunbar P (1974d) Mammals and birds of the Simien Mountains National Park. Walia Eth Wildl Nat Hist Soc Addis Ababa 5:4/5

Dunbar R, Dunbar P (1975) Social dynamics of Gelada baboons. Contributions to primatology, vol VI. Karger, Basel; 157 pp.

Ellenberg H (1978) Zur Populationsökologie des Rehes (*Capreolus, capreolus* L., Cervidae) in Mitteleuropa. Spixiana Suppl 2:211

Estes RD (1966) Behaviour and life history of the wildebeest (*Connochaetes taurinus* Burchell). Nature (London) 212:999-1000

Estes RD (1974) Social Organization of the African Bovidae. In: Geist V, Walther F (eds) The Behaviour of ungulates and its Relation to Management; IUCN Pub No 24 Morges, 166-205

Farrand WR (1971) Late quaternary paleoclimates of the eastern mediterranean area. In: Turekian KK (ed) Late cenozoic glacial ages. Yale University Press: 529-564

Fedigan LM (1972) Roles and activities of male geladas (*Theropithecus gelada*). Behaviour 41:82-90

Ferrar AA, Walker BH (1974) An analysis of Herbivore/Habitat relationship in Kyle National Park, Rhodesia. J S Afr Wildl Manage Ass 4 (3):137-147

Flury P (1963) Appr: Der Steinbock in der Antike. Swiss Foundation for Alpine Research, Zürich, unpublished

Frei E (1978) Andepts in some high mountains of east Africa. Geoderma 21:119-131

Gasse F, Rognon P, Street FA (1980) Quaternary history of the Afar and Ethiopian Rift Lakes. In: Williams MAJ, Faure H (eds) The Sahara and the Nile. Balkema, Rotterdam, in press

Geist V (1971a) Mountain sheep, a study in behaviour and evolution. Wildl Behav Ecol Ser. University Chicago Press, Chicago London, 383 pp.

Geist V (1971b) Tradition und Arterhaltung bei Wildschaf und Elch; n + m, Naturwissenschaft und Medizin, 8 Jg, Nr 37, Boehringer, Mannheim, pp 25-35

Geist V (1974a) On the relationship of ecology and behaviour in the evolution of ungulates: theoretical considerations. In: Geist V, Walther F (eds) The behaviour of ungulates and its relation to management. New Ser No 24, IUCN Publ, Morges

Geist V (1974b) On the relationship of social evolution and ecology in ungulates. Am Zool 14 (1):205-220

Geist V (1977) A comparison of social adaptations in relation to ecology in Gallinaceous bird and ungulate societies. Ann Rev Ecol Syst 8:193-207

Gentry AW (1970) The bovidae (Mammalia) of the Fort Ternan Fossil fauna. In: Leakey LSB, Savage RJG (eds) Fossil vertebrates of Africa. Academic Press, London New York; 243-323

Gerster G (1973) Hoch-Semien, das Dach Aethiopiens. Schweizer Unterstützung für einen afrikanischen Nationalpark. Neue Züricher Zeit (452); 30.9.1973, pp 55-57

Gerster G (1974) Aethiopien, das Dach Afrikas. Atlantis, Zürich Freiburg, pp 211-222

Gessner O Die Gift- und Arzneipflanzen von Mitteleuropa. C Winter, Heidelberg; p. 24

Gray AP (1972) Mammalian hybrids; a check-list with bibliography, 2nd ed. Commonwealth Agricultural Bureaux Farnham Royal, Slough; pp 128-133

Grimwood IR (1965) Ethiopia, conservation of natural resources. Mimeogr, UNESCO, Paris

Grove AT, Street FA, Goodie AS (1975) Former lake levels and climate change in the rift valley of Southern Ethiopia. Geogr J 141:177-202

Grubb P, Jewell PA (1966) Social grouping and home range in feral Soay sheep. Symp Zool Soc London 18:179-210

Haltenorth T (1963) Klassifikation der Säugetiere: Artiodacyla I (18):1-167, Handb Zool

Hamilton AC (1972) The interpretation of pollen diagrams from highland Uganda. In: Zinderen Bakker EM (ed) Palaeoecology of Africa, the surrounding islands and Antarctica, vol VII, Balkema, Cape Town; 45-117

Hamilton AC (1974) The history of the vegetation. In: Lind EM, Morrison MES (eds) East African vegetation. Longman, London; pp 188-209

Hamilton AC (1977) An upper pleistocene pollen diagram from Highland Ethiopia. Abstr Xth INQUA Congr, Birmingham

Hamilton AC, Perrott A (1978) Date of deglacierisation of Mount Elgon. Nature (London) 273:49

Harper F (1945) Extinct and vanishing mammals of the old world. Spec Publ, vol 12. New York Zool Park, New York, p 623

Hastenraths S (1974) Glaziale und periglaziale Formbildung in Hoch-Semyen, Nord-Aethiopien Erdkde 28:176-186

Hausfater G (1975) Dominance and reproduction in baboons (*Papio cynocephalus*). A quantitative analysis. Contrib Primatol, vol VII Karger, Basel, 150 pp.

Hedberg O (1955) Altitudinal zonation of the vegetation on the east Africa mountains. Proc Linn Soc London 165: Session 1952-53, 134-136

Hedberg O (1964) Features of Afroalpine plant ecology. Acta Phytogeogr Suecica 49. Almqvist Wiksells, Uppsala; 144 pp.

Hendrichs H, Hendrichs U (1971) Freilanduntersuchungen zur Oekologie und Ethologie der Zwergantilope Madoqua (*Rhynchotragus*) Kirki, Günter 1880. Piper-Verlag, München, "Dikdik und Elefanten", pp 9-75

Hensel H (1955) Mensch und warmblütige Tiere. Third part. In: Temperatur und Leben. Springer, Berlin Heidelberg New York

Heptner VG, Nasimovic AA, Bannikov AG (1966) Die Säugetiere der Sowjetunion. VEB G Fischer, Jena; 939 pp.

Herre W (1958) Abstammung und Domestikation der Haustiere. In: Hammond J, Johansson I, Haring F (eds) Handbuch der Tierzüchtung, Bd 1. Paul Parey, Hamburg; pp 1-58

Herre W, Röhrs M (1955) Über die Formenmannigfaltigkeit des Gehörns der Caprini Simpson 1945. Zool Gart 22:85-110

Hofmann A, Nievergelt B (1972) Das jahreszeitliche Verteilungsmuster und der Aesungsdruck von Alpensteinbock, Gemse, Rothirsch und Reh in einem begrenzten Gebiet im Oberengadin. Z. Jagdwiss 18:185-212

Hurni H (1975a) Bodenerosion in Semien-Aethiopien. Geogr Helv 30:157-168

Hurni H (1975b) Simen Mountains National Park, Park Warden's report concerning the months of March, April and May 1975. Eth Wildl Conserv Org. WWF Morges, Addis Abeba, polycopied; 10 pp.

Hurni H (1976) Simien Mountains National Park, Warden's report January to March 1976, incl. Warden's report about his trip to remote Walia areas. Wildl Conserv Dep. WWF Morges, Addis Abeba, polycopied

Hurni H (1978) Soil erosion forms in the Simen mountains-Ethiopia (with map 1:25 000). In: Messerli B, Aerni K (eds) Simen mountains - Ethiopia, vol I. Cartography and its application for geographical and ecological problems. Geographica bernensia G8; pp 93-100

Hurni H (1980a) Die Dynamik der Höhenstufung im Semien-Gebirge von der letzten Kaltzeit bis zur Gegenwart (with 3 maps in English). In: Das Hochgebirge von Semien-Aethiopien; vol II. Geographica Bernensia, Bern, in preparation

Hurni H (1980b) Soil erosion processes and soil conservation design in the Simen mountains (Ethiopia). In preparation

Hurni H, Stähli P (1980) Beiträge zum Klima des Semien-Gebirges. In: Das Hochgebirge von Semien-Aethiopien; vol II. Geographica Bernensia, Bern, in preparation

Huwyler E (1977) Der Steinbock in der Vorstellungswelt des Mungan-Tals (Nordost-Afghanistan). Lizentiatsarbeit, Eth Seminar, Univ Basel

Iwamoto T (1979) Feeding ecology, Chap. 12. In: Kawai M (ed) Ecological and Sociological studies of Gelada baboons. Contrib Primatol vol 16, Karger and Kodanska, pp 279-330

Jarman PJ (1974) The social organisation of Antilope in relation to their ecology. Behaviour 58 (3,4):215-267

Jones PJ, Ward P (1976) The level of reserve protein as the proximate factor controlling the timing of breeding and clutch-size in the red-billed quelea (*Quelea quelea*). Ibis 118:547-574

Kaczmarski F (1966) Bioenergetics of pregnancy and lactation in the bank vole. Acta Theriol 11:409-417

Karrer P (1959) Lehrbuch der organischen Chemie. 13 Aufl. G Thieme, Stuttgart, pp 907/908

Kawai M (ed) (1979) Ecological and sociological studies of Gelada baboons. Contrib Primatol, vol 16. Karger and Kodanska; 344 pp.

Kawai M, Iwamoto T (1979) Nomadism and activities, Chap. 11. In: Kawai M (ed) Ecological and sociological studies of Gelada baboons. Contrib Primatol, vol 16. Karger and Kodanska; pp 251-278

Kendall RL (1969) An ecological history of the Lake Victoria Basin. Ecol Monogr 39:121-176

Kesper K-D (1953) Phylogenetische und entwicklungsgeschichtliche Studien an den Gattungen Ovis und Capra. Diss Univ Kiel

Kienholz H, Messerli B (eds) (1974) Sahara, Bericht über die Sahara-Exkursion des Geographischen Institutes der Universität Bern. Funk Helio, Bern; 256 pp.

Kleiber M (1961) The fire of life. An introduction to animal energetics. Wiley, New York London, 453 pp.

Klingel H (1967) Soziale Organisation und Verhalten freilebender Steppenzebras, 2. Tierpsychologie 24 (5):580-624

Klötzli F (1972) Walia ibex in Simien Mountains National Park; Ecological survey on habitat conservation. World Wildl Yearb 1971-72:81-84

Klötzli F (1975a) Zur Waldfähigkeit der Gebirgssteppen Hoch-Semiens (Nordäthiopien). Beitr Naturkd Forsch Suedwestdtschl 34:131-147

Klötzli F (1975b) Am seltensten ist der Semienfuchs. Kosmos 4:148-156

Klötzli F (1975c) Simen - a recent review of its problems; Walia. J Eth Wildl Nat Hist Soc 6:18/19

Klötzli F (1977) Wald und Vieh im Gebirgsland Aethiopiens. In Tüxen R (ed) Vegetation und Fauna. J Cramer, Vaduz, pp 499-512

Kock D (1971) Zur Verbreitung von Mähnenschaf und Steinbock im Nilgebiet. Ein Beitrag zur Zoogeographie Nordost-Afrikas. Säugetierkdl Mitt 19 Jg 1:28-39

Krämer A (1969) Soziale Organisation und Sozialverhalten einer Gemspopulation (*Rupicapra rupicapra* L.) der Alpen. Tierpsychologie 26:889-964

Kruuk H (1967) Competition for food between vultures in East Africa. Ardea 55:171-193

Kruuk H (1978) Spatial organization and territorial behaviour of the European badger Meles meles. J Zool London 184:1-19

Kummer H (1974) Distribution of interindividual distances in patas monkeys and Gelada baboons. Folia Primate 21:153-160

Lamprey HF (1963) Ecological separation of the large mammal species in the Tarangire game reserve, Tanganyika. East Afr Wildl 1:63-92

Leuthold W (1977) African ungulates, a comparative review of their ethology and behavioural ecology zoophysiology and ecology, vol VIII. Springer, Berlin Heidelberg New York

Lilyestrom WE (1974) Birds of the Simen Highlands. Walia J Eth Wildl Nat Hist Soc 5:2/3

Lind EM, Morrison MES (1974) East African vegetation. Longman, London; 257 pp.

Livingstone DA (1967) Postglacial vegetation of the Ruwenzori Mountains in Equatorial Africa. Ecol Monogr 37:25-52

Livingstone DA (1971) A 22 000-year pollen record from the Plateau of Zambia. Limnol Oceanogr 16:349-356

Livingstone DA (1975) Late quaternary climatic change in Africa. Ann Rev Ecol Syst 6:249-280

Lydekker R (1913) Catalogue of the ungulate mammals in the British Museum (Nat Hist), vol I. Johnson Reprint, New York London; p 155

Mac Arthur RH (1972) Geographical ecology. Harper and Row, New York, 269 pp.

Mac Arthur RH, Wilson EO (1967) The theory of island biography. Princeton Univ Press; 203 pp.

Marler P (1968) Colobus guereza: territoriality and group composition. Science 163:93-95

Maydon HC (1925) Simen, its heights and abysses; a record of travel and sport in Abyssinia. HF Witherby, London; 235 pp.

McMillan JF (1953) Measures of Association between Moose and Elk on feeding grounds. J Wildl Manage 17:162-166

Mertens R (1949) Eduard Rüppell, Leben und Werk eines Forschungsreisenden. W Kramer, Frankfurt aM, 338 pp.

Mesfin Wolde Mariam (1970) An atlas of Ethiopia. Revised ed, Il Poligrafico, Asmara, 84 pp.

Mesfin Wolde Mariam (1972) An introductory geography of Ethiopia. Berhanena Selam Printing Press, Addis Abeba, 215 pp.

Messerli B (1975) Formen und Formungsprozesse in den Hochgebirgen Aethiopiens. Tagungsber Wiss Abh 40. Dtsch. Geographentag Innsbruck. F Steiner, Wiesbaden; pp 389-395

Moen AN (1973) Wildlife ecology. An analytical approach. WH Freeman, San Francisco, 458 pp.

Mori U (1979) Reproductive behaviour. In: Kawai M (ed) Ecological and sociological studies of Gelada baboons. Contrib Primatol, vol 16. Karger and Kodanska; pp 183-197

Müller JP (1972) Report on an overall census of Walia ibex (Capra walie) in the Simen Mountains National Park in February 1972, Mimeograph. Eth Wildl Dep. WWF Morges, Addis Abeba, Pro Semien Zürich

Müller JP (1973a) Provision of a warden for Simien Mountains National Park. World Wildl Yearb 1972-73:78-84

Müller JP (1973b) Simien Mountains National Park: Outlook on management problems and projects. Mimeograph, Chur; 16 pp.

Müller JP (1974) Project 666. Provision of warden for Simien Mountains National Park. World Wildl Yearb 1972-73:78-84

Müller JP (1977) Populationsökologie von Arvicanthis abyssinicus in der Grassteppe des Semien Mountains National Park (Aethiopien). Z Säugetierkunde 42:145-172

Murphy D (1968) In Ethiopia with a mule. J Murray Lewis Reprints Ltd, London; 281 pp.

Myers N (1973) Leopard and cheetah in Ethiopia. Oryx XII:197-205

Naegeli R (1978) Debark (Simen)- a market town in the highland of Ethiopia (with 2 maps). In: Messerli B, Aerni K (eds) Simen mountains- Ethiopia, vol I. Cartography and its application for geographical and ecological problems. Geographica bernensia G8, 73-91

Nicol CW (1972) From the roof of Africa. AA Knopf, New York; 362 pp.

Nievergelt B (1966a) Der Alpensteinbock (Capra ibex L) in einem Lebensraum. Ein ökologischer Vergleich. Mammalia depicta. P Parey, Hamburg Berlin; 85 pp.

Nievergelt B (1966b) Unterschiede in der Setzzeit beim Alpensteinbock (Capra ibex L). Rev Suisse Zool 73 (26):446-454

Nievergelt B (1967) Die Zusammensetzung der Gruppen beim Alpensteinbock, 2. Säugetierkunde 32 (3):129-144

Nievergelt B (1968) Die Steinbock-Kolonie am Piz Albris. Geschützte Natur. Ziemsen-Verlag, Wittenberg Lutherstadt; pp 155-158

Nievergelt B (1969a) Simien, eine bedrohte Berglandschaft in Aethiopien. Berge der Welt, Bd 17. Schweiz Stift Alp Forsch, Zürich (engl transl); pp 133-138

Nievergelt B (1969b) Le bouquetin d'Ethiopie gravement menacé. Biol Conserv 1 (4):334-335

Nievergelt B (1969c) Ecological survey in the Simen mountains, Ethiopia. World Wildl Fund Yearb 1968:95-97

Nievergelt B (1970a) The Walia ibex (Capra walie) of Ethiopia and its annual reproductive cycle. Trans 9th Int Congr Game Biol, Moscow; pp 868-871

Nievergelt B (1970b) Der äthiopische Steinbock ist in Gefahr. Panda Jg 3 H 1:20-24

Nievergelt B (1971) A method for estimating the increase or decrease in population of Walia ibex in the Semien Mountains National Park. Mimeograph. Zool Inst, Zürich Univ; 15 pp.

Nievergelt B (1972a) Der Walia-Steinbock (Capra walie). Bestand und Fortpflanzung einer bedrohten Art. Una vita per la natura, Camerino; pp 203-209

Nievergelt B (1972b) The situation in the Semien Mountains National Park in 1971 compared with 1966 and 1968/69. Mimeogr Rep Eth Wildl Conserv Org WWF, Zürich Univ; 15 p.

Nievergelt B (1973) Erhaltung der Lebensgrundlagen - ein dringliches Hilfsprojekt in den Semien-Bergen in Aethiopien. Mimeograph. Pro Semien, Zürich; 16 p.

Nievergelt B (1974) A comparison of rutting behaviour and grouping in the Ethiopian and Alpine ibex. In: Geist V, Walther F (eds) The behaviour of ungulates and its relation to management. IUCN Publ No 24, Morges, 324-340

Nievergelt B (1978) Die Knoten am Bockgehörn von Bezoarziege und Alpensteinbock (*Capra a aegagrus* und *Capra ibex*). Säugetierkunde 43 (3):187-190

Ohsawa H (1979) Population, Part I. In: Kawai M (ed) Ecological and sociological studies of Gelada baboons. Contrib Primatol, vol 16. Karger and Kodanska; pp 1-80

Orians G (1969) On the evolution of mating systems in birds and mammals. Am Nat 103:589-603

Parisi B (1925) Sulla Capra Walie, Rüppell; Atti Soc Ital Sci Nat Mus Civ Stor Nat Milano 64:110-118

Pfeffer P (1967) Le mouflon de Corse. Suppl Tome 31, Mammalia, Paris, 262 pp.

Preston FW (1948) The commonness and rarity of species. Ecology 29:254-283

Qvortrup SA, Blankenship LH (1974) Food habits of Klipspringer. East Afr Wildl J 12:79-80

Richardson JL, Richardson AE (1972) History of an African rift lake and its climate implications. Ecol Monogr 42:499-534

Ricklefs RE (1979) Ecology, Second Edition. T. Nelson Ltd, Sunbury-on-Thames, 966 pp.

Rüppell E (1835) Neue Wirbelthiere zu der Fauna von Abyssinien gehörig. Säugethiere 40 S, 14 Taf. Frankfurt aM

Rüppell E (1838) Reise in Abyssinien, 2 vols. S Schmerber, Frankfurt aM, 435 and 448 pp.

Schaerer O (1977) Standortwahl, Tagesaktivität und Verbandstruktur in einem Bockrudel des Alpensteinbockes (*Capra ibex* L.). Diplomarbeit Univ Zürich

Schaerer O (1979) A bibliography on nature and man of the Simen Mountains (Ethiopia). Ethol Wildforsch, Zürich Univ; 147 pp.

Schaller GB (1967) The deer and the tiger: a study of wildlife in India. University of Chicago Press; 370 pp.

Schaller GB (1972) The Serengeti lion. A study of predator-prey relations. University of Chicago Press; 480 pp.

Schaller GB (1977) Mountain monarchs, wild sheep and goats of the Himalaya. University of Chicago Press; 425 pp.

Schloeth R, Burckhardt D (1961) Die Wanderungen des Rotwildes *Cervus elaphus* L, im Gebiet des Schweizerischen Nationalparkes. Rev Suisse Zool 68:145-156

Schultze-Westrum T (1963) Die Wildziegen der ägäischen Inseln. Säugetierkundl Mitt 11:145-182

Simpson GG (1945) The principles of classification and a classification of mammals. Bull Am Mus Nat Hist 85:350 pp.

Soulé ME, Wilcox BA, Holtby C (1979) Benign neglect: A model of faunal collapse in the game reserves of East Africa. Biol Conserv 15:259-272

Stähli P (1977) Semien - Gefährdete Landschaft im äthiopischen Bergland. Berner Geogr Mitt 1976:23-24

Stähli P (1978) Changes in settlement and land use in Simen, Ethiopia, especially from 1954 to 1975. In: Messerli B, Aerni K (eds) Cartography and its application for geographical and ecological problems Simen Mountains - Ethiopia, vol I. Geogr Inst Univ Bern; pp 33-72

Stähli P, Zurbuchen M (1978) Two topographic maps 1:25 000 of Simen, Ethiopia. In: Messerli B, Aerni K (eds) Simen Mountains-Ethiopia, vol I. Cartography and its application for geographical and ecological problems. Geographica bernensia G8; pp 11-31

Street FA (1980a) Late quaternary precipitation estimates for the Ziway-Shala-Basin, Southern Ethiopia. Erdkunde, in press

Street FA (1980b) Chronology of late pleistocene and holocene lake-level fluctuations, Ziway-Shala Basin, Ethiopia. Proc VIII Panafr Congr Prehist Quat Stud, Nairobi (held in 1977), in press

Street FA, Grove AT (1976) Environmental and climatic implications of late quaternary lake-level fluctuations in Africa. Nature, London 261:385-390

Talbot LM (1966) Wild animals as a source of food. US Dep Interior, Spec Sci Rep Wildl 98:16

Talbot L, Talbot M (1963) The wildebeest in western Masailand, East Africa. Wildl Monogr 12. The Wildlife Society; 88 pp.

Terborgh J (1975) Faunal equilibra and the design of wildlife preserves. In: Golley F, Medina E (eds) Trop Ecol Syst. Trends in terrestrial and aquatic Research. Springer, Berlin Heidelberg New York, pp 369-380

Tewoldeberhan Gebregziabher (1974) Soil conservation in Ethiopia, II. Some problems of afforestation and forest management. Walia J Eth Wildl Nat Hist Soc 5:12-14

Tinley KL (1969) Dik dik Madoqua kirki in South-west Africa: notes on distribution, ecology and behaviour. Madoqua 1:7-33

Türcke F, Schmincke S (1965) Das Muffelwild, Naturgeschichte, Hege und Jagd. P Parey, Hamburg; 193 pp.

Urban EK, Brown LH (1971) A checklist of the birds of Ethiopia. Haile Sellasie I University Press, Addis Ababa; 143 pp.

Veitch RS (ed) (1972) Ethiopian endeavour. The John Hunt Exploration Group of endeavour training expedition to the High Simiens of Ethiopia. JW Thompson, London; 109 pp.

Verfaillie M (1978 The ericaceous belt of the Semien Mountains National Park, Ethiopia. Biol Jaarb 46:210-223

Vollmar F (1969) The conservation situation in Ethiopia. World Wildl Fund Yearb 1968:25-31

Vos De A. (1975) Africa, the devastated continent. W Junk, The Hague; 236 pp.

Voser-Huber ML, Nievergelt B (1975) Das Futterwahlverhalten des Rehes in einem voralpinen Revier. Z Jagdwiss 21:197-215

Walther F (1968) Verhalten der Gazellen. Neue Brehm-Bücherei. A Ziemsen, Wittenberg Lutherstadt; 144 pp.

Weiner J (1977) Energy metabolism of the roe deer. Acta Theriol 22:3-24

Werdecker J (1958) Untersuchungen in Hochsemien. Mitt Geogr Ges Wien 100:58-69

Werdecker J (1961) Geographische Forschungen in Nordäthiopien. In: Asmus W, Ruppert JP (eds) Erziehung als Beruf und Wissenschaft. M Diesterweg, Frankfurt, pp 150-157

Werdecker J (1967) Karte Hoch-Semyén (Aethiopien), 1:50000, Deutsche Forschungs- gemeinschaft, München (Hrsg). Beil II Erdkunde 22:1

Wildlife Conservation Regulations (1972) Legal notice No. 416 of 1972. Negarit Gaz 31 (7): 19-1-1972, 35-52

Williams MAJ, Street FA, Dakin FM (1978) Fossil periglacial deposits in the Simen Highlands, Ethiopia. Erdkunde 32:40-46

Wilson EO (1975) Sociobiology, the new synthesis. Belknap Press of Harvard Univ Press; 697 pp.

Wilson EO, Willis EO (1975) Applied biogeography. In: Cody ML, Diamond JM (eds) Ecology and evolution of communities. Harvard University Press; pp 522-534

Winiger M (1976) Das geplante Land- und Forstwirtschaftliche Zentrum. Jahresber 1974/75. Pro Semien, c/o Zurich University, polycopied, pp 15-18

Zeuner FE (1945) The pleistocene period, its climate, chronology and faunal successions. Ray Society, London; 322 pp.

Zinderen Bakker Van EM (1962) Botanical evidence for quaternary climates in Africa. Ann Cape Prov Mus II. South Africa: 16-31

Zinderen Bakker Van EM (1964) A pollen diagram from Equatorial Africa Cherangani, Kenya. Geol Mijnbouw 43:123-128

Zinderen Bakker Van EM (1967) Upper pleistocene and holocene stratigraphy and ecology on the basis of vegetation changes in Sub-Saharan Africa. In: Bishop WW, Clark JD (eds) Background to evolution in Africa. Univ of Chicago Press; p 125

Zinderen Bakker Van EM (1969) Intimations on quaternary palaeoecology of Africa. Acta Bot Neerl 18 (1):230-239

Zingg R (1980) Aehnlichkeitsbeziehungen zwischen den Vertretern der Gattung Capra aufgrund gehörnmorphologischer Strukturen. Diplomarbeit Univ Zürich

Zittel KA (1891-1893) Palaeozoologie. Handb Palaeontologie, Bd IV. Vertebrata, München Leipzig; p 422

Subject Index

Page numbers in *italics* refer to Figures
Page numbers in **bold face** refer to Tables

Zoophysiology

(formerly **Zoophysiology and Ecology**)

Coordinating Editor: **D.S.Farner**
Editors: **W.S.Hoar, B.Hoelldobler, K.Johansen, H.Langer, M.Lindauer**

A Selection

Volume 8
W.Leuthold

African Ungulates

A Comparative Review of their Ethology and Behavioral Ecology

1977. 55 figures, 7 tables. XIII, 307 pages
ISBN 3-540-07951-3

"...Dr. Leuthold displays a masterly command of his subject... The work is basically a review of published knowledge with an original approach, enlivened by the author's interpretations and based on his intimate first-hand knowledge of the subject. The first chapter, on the application of ethological knowledge to wildlife management, covers an important area... a wealth of references is given so that the chapter provides a useful guide to the literature. The illustrations are good and well chosen to demonstrate points made verbally and not first to embellish the text. The book will provide **excellent background reading for undergraduates and research students as well as for anyone seriously interested in African wildlife.** On the whole, **it can be thoroughly recommended.**" *J. Applied Ecology*

Volume 9
E.B.Edney

Water Balance in Land Arthropods

1977. 109 figures, 36 tables. XII, 282 pages
ISBN 3-540-08084-8

"...Dr. Edney has provided a wealth of organized information on prior work and ideas for needed research, **all of which make the book a bargain.** The volume should prove useful, not only to those who work in arthropod water relations (it is a must for them), but also to those of us interested in invertebrate and general ecology, entomology, comparative physiology, and biophysics." *AWRA Water Resources Bull*

"...The virtue of the work is that it reviews most comprehensively, and it reports accurately in clear and simple style, the extensive literature on cuticular permeability..." *Nature*

Volume 10
H.-U.Thiele

Carabid Beetles in Their Environments

A Study on Habitat Selection by Adaptations in Physiology and Behaviour

Translated from the German by J. Wieser
1977. 152 figures, 58 tables. XVII, 369 pages
ISBN 3-540-08306-5

"...Because the book is comparative both in method and interpretation, it is a contribution to systematics as well as to ecology ...**a fine synthesis of current knowledge** of homeostatic aspects of ecological relationships of carabids, and it is a fitting tribute to the man to whom it is dedicated: Carl H. Lindroth, who was instrumental in formulating the approaches and techniques that are commonly used in ecological research on these fine beetles. The material is **well organized** and the text is **easily readable, thanks to the clarity of thought and expression** of the author and to the skill of an able translator." *Science*

Volume 11
M.H.A.Keenleyside

Diversity and Adaptation in Fish Behaviour

1979. 67 figures, 15 tables. XIII, 208 pages
ISBN 3-540-09587-X

"...it is important as the first serious attempt by a senior researcher to produce an overview of the discipline. Previous works have all been symposium volumes or collections of papers haphazardly assembled, and Keenleyside has produced a volume that is of substantially greater value than these. In clearly perceiving that the unique and valuable features of fish behavior are its diversity of form and circumstance, he has charted a course that future authors would be wise to follow.
The book is well produced, well written, and easy to read. The illustrations are clear and straightforward." *Science*

Springer-Verlag
Berlin
Heidelberg
New York

Ecological Studies

Analysis and Synthesis

Editors: **W. D. Billings, F. Golley, O. L. Lange, J. S. Olson**

A Selection

Volume 34
Agriculture in Semi-Arid Environments

Editors: **A. E. Hall, G. H. Cannell, H. W. Lawton**
With contributions by numerous experts
1979. 47 figures, 22 tables. XVI, 340 pages
ISBN 3-540-09414-8

Consisting of reviews and analyse of information from researchers in the United States, Africa, and Australia, **Agriculture in Semi-Arid Environments** focuses on dryland farming techniques for areas of the world where rainfall is limited and water is not freely available for irrgation. This volume examines dryland farming systems that have been used in the past and evaluates those systems as possible approaches or models for contemporary agricultural development.
Several chapters explore the specific problems of West Africa particularly the agroclimatology, plant diseases, and nematode pests of the Sahelian and Sudanian semi-arid zones. Finally, the interaction between cultivation and livestock production are discussed, revealing the need for an integrated multidisciplinary approach to agricultural development in semi-arid environments.

Volume 35
J. A. C. Fortescue
Environmental Geochemistry

A Holistic Approach

1980. 131 figures. XVII, 347 pages
ISBN 3-540-90454-9

This volume introduces the current status of environmental geochemistry and discusses the geological and environmental factors surrounding the synthesis and decomposition of many environmental materials. This includes examinations of element abundance, biogeochemical cycling of elements and compounds, and the geochemical classification of landscapes. Dr. Fortescue's holistic approach encompasses the study of all components at or near the earth's surface. Rocks, soils, and vegetation are treated as a single working unit.
Environmental Geochemistry is valuable to all advanced students and researchers in the environmentally related fields of pollution studies, land use, and environmental health. Workers in ecology, geography, soil science, forestry, and epidemiology will also find this book indispensable.

Volume 36
C. B. Osmond, O. Björkman, D. J. Anderson
Physiological Processes in Plant Ecology

Toward a Synthesis with **Atriplex**

1980. 194 figures, 76 tables. XI, 468 pages
ISBN 3-540-10060-1

This book provides a uniquely elegant framework for the evaluation and integration of physiological plant ecology studies. It illustrates the continuum of research in this field with examples drawn largely from the widespread genus *Atriplex*, whose members have been ecologically successful in a variety of habitats.
Particular emphasis is placed on genealogical differentiation and evolutionary relationships, community ecology, reproductive strategy and seedling establishment, and detailed physiological processes such as ionic uptake, water relations and photosynthesis. Researchers in plant physiology, biochemistry and ecology will welcome this distinctive approach to a better understanding of plant adaptation, both in terms of plant performance in response to environment and in terms of persistence in different habitats.

Volume 37
H. Jenny
The Soil Resource

Origin and Behavior

1980. 191 figures. XX, 377 pages
ISBN 3-540-90543-X

Any examination of the nature of soil exposes a bewildering array of complex relationships between the chemical, biological, and physical components. **The Soil Resource** provides in a single volume the necessary scientific background for a comprehensive understanding of soils, the role they play in our environment, and their place in agriculture and forestry. It treats soils as part of the land ecosystem, not, as is usually the case, as an isolated element, and as structured bodies composed of both biotic and abiotic components.
Part A examines soil processes in terms of atoms, molecules, colloids, enzymes, organisms, and their interactions. **Part B** discusses soil variation through time and space. Climatic, organismic, and topographical models are discussed, as are soil origins and age sequences.

Springer-Verlag
Berlin Heidelberg New York